Brain Microdissection Techniques

IBRO HANDBOOK SERIES: METHODS IN THE NEUROSCIENCES

General Editor: **A. D. Smith**
Department of Pharmacology
University of Oxford

Volume 1: **Tracing Neural Connections with Horseradish Peroxidase**
Edited by M-Marsel Mesulam

0 471 10028 5	280 pages	(cloth)	1982
0 471 10029 3	280 pages	(paper)	1982

Volume 2: **Brain Microdissection Techniques**
Edited by A. C. Cuello

Volumes in Preparation

Receptor Binding Methods
N. Birdsall and E. C. Hulme

Immunohistochemistry
Edited by A. C. Cuello

Histochemical and Ultrastructural Identification of Monoamine Neurons
Edited by J. B. Furness and M. Costa

Chemical Lesions
Edited by G. Jaim-Etcheverry

Intracellular Perfusion of Excitable Cells
Edited by P. G. Kostyuk and O. A. Krishtal

Methods for Neuronal Recording in Conscious Animals
R. Lemon

Brain Microdissection Techniques

Edited by
A. C. Cuello
Departments of Pharmacology and Human Anatomy,
University of Oxford, UK

A Wiley–Interscience Publication

JOHN WILEY & SONS
Chichester · New York · Brisbane · Toronto · Singapore

Library of Congress Cataloging in Publication Data:
Main entry under title:
Brain microdissection techniques.
(IBRO handbook series)
'A Wiley–Interscience publication.'
Includes index.
1. Brain—Anatomy. 2. Dissection. 3. Dissecting microscope. I. Cuello, A. C. II. Series.
[DNLM: 1. Brain—Anatomy and histology.
2. Dissection—Methods. WL 300 B8136]
QM455.B64 1983 611'.81'078 82-111099

ISBN 0 471 10523 6 (cloth)
ISBN 0 471 90019 2 (paper)

British Library Cataloguing in Publication Data:
Brain microdissection techniques.—
(IBRO handbook series: methods in neurosciences; v. 2)
1. Brain 2. Dissection
I. Cuello, A. C. II. Series
611 QM453

ISBN 0 471 10523 6 (cloth)
ISBN 0 471 90019 2 (paper)

Text set in 10/12 pt Linotron 202 Times, printed and bound in Great Britain at The Pitman Press, Bath

Contributors

S.-M. Aquilonius	*Dept. of Neurology, University Hospital, S-750 14 Uppsala, Sweden*
E. D. Bird	*Brain Tissue Resource Center, Ralph Lowell Laboratories, McLean Hospital, Harvard Medical School, Belmont, Massachusetts 02178, USA*
M. J. Brownstein	*Laboratory of Clinical Science, National Institute of Mental Health, Bethesda, Maryland 20205, USA*
S. Carson	*Neuropharmacology/Neuroanatomy Group, Department of Pharmacology, University of Oxford, UK*
A. C. Cuello	*Neuropharmacology/Neuroanatomy Group, Departments of Pharmacology and Human Anatomy, University of Oxford, UK*
S.-Å. Eckernäs	*Dept. of Neurology, University Hospital, S-750 14 Uppsala, Sweden*
P.-G. Gillberg	*Dept. of Neurology, University Hospital, S-750 14 Uppsala, Sweden*
I. Kanazawa	*Dept. of Neurology, Institute of Clinical Medicine, University of Tsukuba, Ibaraki-ken 305, Japan*
M. Palkovits	*1st Department of Anatomy, Semmelweis University Medical School, Budapest, H-1450, Hungary*
T. J. Stevens	*Brain Tissue Resource Center, Ralph Lowell Laboratories, McLean Hospital, Harvard Medical School, Belmont, Massachusetts 02178, USA*

Contents

Series Preface viii

Preface x

1. Microdissection of Brain Areas by the Punch Technique
Miklós Palkovits and Michael J. Brownstein 1

2. Microdissection of Fresh Rat Brain Tissue Slices
A. Claudio Cuello and Susan Carson 37

3. Grid Microdissection of Human Brain Areas
Ichiro Kanazawa 127

4. Large Section Cryomicrotomy in Human Neuroanatomy and Neurochemistry
Sten-Magnus Aquilonius, Sven-Åke Eckernäs, and Per-Göran Gillberg 155

5. Microdissections of Nuclear Areas from Human Brain Slices
Thomas J. Stevens and Edward D. Bird 171

Subject Index 185

Preface for IBRO Handbook Series

During the last fifty years there have been two changes in the way in which scientists have studied the nervous system. First of all, the traditional and largely independent major scientific disciplines of physics, chemistry, physiology, pharmacology, and pathology gave rise to the more specialized subdisciplines of neurophysiology, neurochemistry, neuropharmacology, etc., and the science of experimental psychology was born. Then, after about another generation, it became clear that a deeper understanding of the brain could not be achieved by separate and unrelated studies in each of these sub-disciplines. Rather, a unified approach was needed in which the specialized methods were applied in a co-ordinated way to solve a particular problem. Indeed, combinations of methods could often yield results not obtainable by the application of any individual technique. Thus, scientists studying the nervous system began to call themselves neurobiologists or neuroscientists because they did not wish to be identified with any particular experimental discipline. Very soon meetings took place (e.g. in 1955 the First International Meeting of Neurobiologists) and organizations (Neuroscience Research Program, MIT) were founded to give formal recognition of this new approach to the study of the brain. In 1958 the decision was taken in Moscow to establish the International Brain Research Organization (IBRO), which became incorporated as an independent organization through a Bill in the Parliament of Canada at Ottawa in 1961.

IBRO now has 2000 members, most of whom hold senior positions in research or teaching, in 52 countries of all political complexions. Through its National Corporate members, many of which are academies of sciences or national societies for neuroscience, the body of neuroscientists reflected in IBRO must be of the order of 15,000. One of the programmes of IBRO, all of which aim to serve the international community of neuroscientists, is the publication programme. IBRO publications include *IBRO News, Neuroscience* and the *IBRO Symposia Series*. With the present *Handbook Series*, IBRO aims to fill a major gap in the world literature. The neuroscientist needs to be able to turn to whichever specialized method that is most suited for the problem he is currently studying. The series on *Methods in the Neurosciences* will help to provide expert advice on exactly how to carry out the experiments, on what difficulties can occur and on the limitations of the method.

It is planned as a continuing series, so that new volumes can be published as and when new methods are developed, tested, and found useful. It is my hope that books in this series will have a significant impact on neuroscience throughout the world, by helping to provide the tools with which the scientist can tackle his problems.

A. David Smith,
Director of Publications,
IBRO,
University Department of Pharmacology,
South Parks Road,
Oxford OX1 3QT,
UK.

Preface

A relatively new approach in neuropharmacology and neurochemistry has been achieved by the use of individually microdissected and anatomically characterized nuclei of the central nervous system. Classical studies in these two fields have dealt with large brain organization such as the 'cortex', 'striatum', 'hypothalamus', etc. The sophisticated biochemical analysis of neurotransmitters, their biosynthetic enzymes and other substances which interest neuroscientists, accelerated the search for their fine microanatomical distribution. The emergence of these new biochemical methodologies, together with the discovery of new neuroactive substances, coincided with the introduction of new techniques to obtain individual nuclei of the central nervous system. Much of the credit for this is due to Miklos Palkovits and his colleagues who, with the aid of the 'micropunch' technique, made this possible. A detailed analysis and description of this procedure is given in Chapter 1 by Drs Palkovits and Brownstein. In Chapter 2 we discuss our approach to the microdissection of fresh brain slices: a method learnt from the originators, Drs Zigmond and Ben-Ari at Dr Leslie Iversen's laboratory.

As some of these methodologies can also be applied to human brain, Drs Aquilonius, Eckernäs, and Gillberg were asked to summarize their application of the 'punch' technique in human brain slices (Chapter 4). The very precise grid method of microdissection of human brain as applied in the laboratory of Dr Kanazawa is given in Chapter 3. This method can, and has been applied to the nervous system of experimental animals.

Finally, the approach of obtaining samples of thick slices of human brain for biochemical studies has been summarized in Chapter 5 by Drs Bird and Stevens. This procedure was worked out after several years of experience in biochemical neuropathology.

It is my hope that this collection of writings will assist my colleagues in the neurosciences to improve or adapt these methods for many potential applications.

A.C.C.

Brain Microdissection Techniques
Edited by A. C. Cuello

CHAPTER 1

Microdissection of Brain Areas by the Punch Technique

MIKLÓS PALKOVITS* and MICHAEL J. BROWNSTEIN†

**1st Department of Anatomy Semmelweis University Medical School Budapest, H-1450, Hungary*
†Laboratory of Clinical Science National Institute of Mental Health Bethesda, Maryland 20205, USA

CONTENTS

1.	Principles and advantages of the micropunch technique	2
2.	Removal of the brain for punching	3
3.	Sectioning the brain	6
	3.1 Manual brain slicing	7
	3.2 Sectioning frozen brains	9
	3.3 Sectioning irradiated brains	11
	3.4 Sectioning fixed tissue	12
4.	Tools for punching	13
	4.1 Magnifying glasses, loupes, microscopes	13
	4.2 Brain sections must be properly illuminated	14
	4.3 Sections must be kept frozen while being dissected	14
5.	Microdissection needles	15
6.	The punch technique	18
	6.1 Practical advice	18
	6.1.1 Training and practice	18
	6.2 Topographical orientation; landmarks	22
	6.3 Micropunch	23
7.	General rules for micropunch dissections	25
8.	Determination of the sample size after the dissection is complete	27
9.	Homogenization of microdissected brain tissue	29
10.	Validating the microdissection method	29
11.	Dissections by one or more people	32
	11.1 One person	32
	11.2 Two people	32
	11.3 Team work	33
	11.4 Optimal protocol for the micropunch dissection	33
12.	Microdissection of the rat brain	34
13.	General comments	35

1. PRINCIPLES AND ADVANTAGES OF THE MICROPUNCH TECHNIQUE

Microdissection techniques for removal of discrete brain areas have become routine in many laboratories because microanalytical techniques allow them to be exploited. The dissection methods that we shall describe are simple and reliable. Furthermore, they can be performed quickly, an advantage when hundreds of samples have to be obtained in the course of an experiment.

While a number of methods for removing cell groups or nuclei from the brain have been detailed, punching tissue samples with a hollow needle from frozen sections cut in a cryostat has become the technique of choice (Palkovits, 1973). Some tried to punch tissue blocks, and subsequently monitored the sampling on sections (Eik-Ness and Brizzee, 1965; Schlumpf *et al.*, 1974). The micropunch technique has also been used to take specimens from unstained fresh brain samples (Jacobowitz, 1974). Methylene blue stained (Zigmond and Ben-Ari, 1976) fresh brain slices have also been applied for microdissection purposes. The staining enables some structures to be visualized more easily.

The punch technique has been applied to a number of problems. It has been used to determine the chemical composition of specific brain cell groups and to reveal the distribution of distinct compounds in the brain. The substances assayed (enzymes, hormones, transmitters, etc.), and the assays used determine how the brain must be removed and how the dissection must be done. The microdissection may be applied—with minor modification—to frozen, fixed, or living brain tissue. Thus, it has been used to sample and orient specific brain regions prior to electron microscopy, and it has been used to remove specific brain regions for tissue cultures. Regardless of the use to which the brain specimens will be put, any dissection method chosen should meet the following requirements:

1. It should be suitable for isolating discrete brain areas. This is not so much a question of technique but of anatomic orientation. Even brain areas with a cross sectional diameter of 0.2 mm may be removed freehand. The removal of brain regions larger than 2 mm involves macrodissection. According to our present knowledge, about 220 cell groups or brain areas may be removed from the rat brain by the punch technique (Palkovits, 1980).
2. The punch technique is suitable for sampling fresh non-fixed, fixed, frozen, or microwave irradiated brain. The tissue specimen removed should be suitable for the use intended. For example, if living tissue is required (for *in vivo* biochemical or pharmacological experiments, studies on tissue cultures) the procedure should ensure the survival of the tissue. If alterations in the brain's chemical constituents *post mortem* are to be avoided, the tissue should be frozen or irradiated.

3. The method should be routinely applicable. The greatest drawback of earlier microdissection procedures was their complexity. As in the course of many experiments hundreds of specimens have to be taken, a simple and rapid microdissecting procedure is a major prerequisite for success. If the samples are properly oriented, hundreds of samples can be removed from frozen sections per hour. It is even possible for one person to remove many thousands of specimens in a day, if necessary.
4. The sampling should be safe, reproducible and easily validated. The conditions for this are well-established, and will be described later. It should be borne in mind that validation of the method is principally the responsibility of the person doing the dissections. Removal of specimens should be strictly limited to those brain areas with which one is familiar. Furthermore, microdissection should not be undertaken if the brain sections are poorly oriented and seem unfamiliar.

The technique developed by us and described below permits, after a rather short period of training, highly rapid sampling. Each step of the punch technique is described in detail. These descriptions are not intended as rules, only as a starting point for learning (and modifying) the technique. One should never forget that applying this method with the attitude 'maybe we are going to find something somewhere' is time consuming and unlikely to be productive.

2. REMOVAL OF THE BRAIN FOR PUNCHING

The technique used to kill the animal and remove the brain is determined by the character and purpose of the experiment. These techniques fall into two groups: (a) Removal of brains without fixation. (b) Removal of brains after intracranial fixation.

Removal of the brain without prior fixation (or freezing) can be employed only if the studies envisioned are not compromised by changes that occur in the tissue in the first two minutes *post mortem*. Certain neurotransmitters, cofactors, etc., are rapidly degraded *post mortem* (e.g., acetylcholine, cyclic AMP) and their *in vivo* levels cannot be estimated in brains that are removed without prior fixation. The stability of any substance to be studied should be established and then the means of killing the animal can be set.

The removal of the brain has to be practiced, 30 seconds ought to be sufficient for removal of a rat brain. Decapitation is the most practical method of killing; cranial damage should be avoided.

The method used to remove the brain will depend on the part of the brain to be microdissected. Ideally, the entire brain should be removed undamaged, but since only seconds are available for this task, this cannot be accomplished invariably. The most frequent defects are the following: (a) The olfactory bulb is retained in the skull. (b) The lower brain stem breaks when

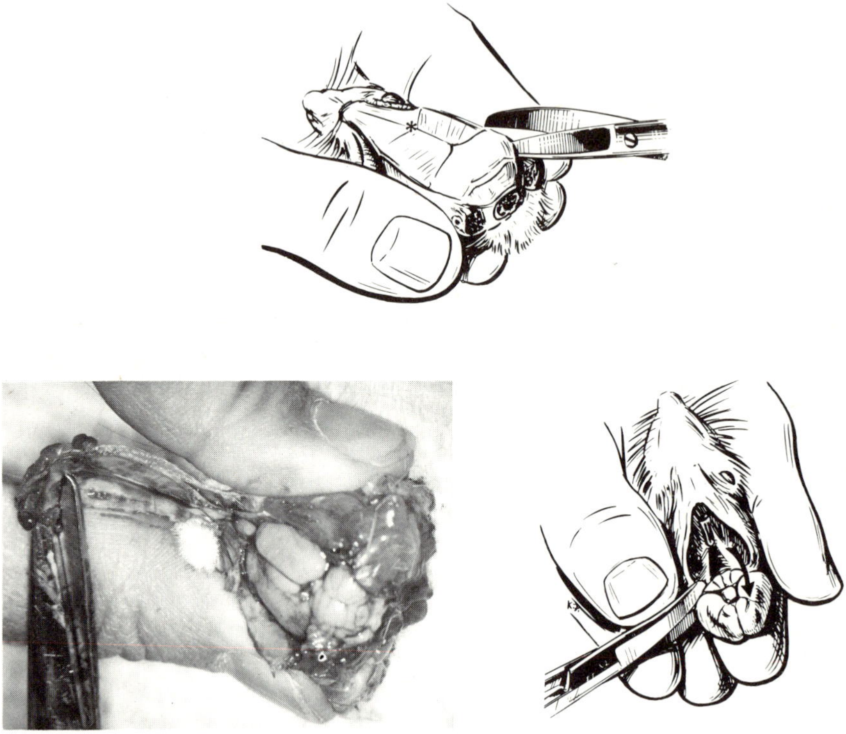

Figure 1 Removal of the rat brain from the calvarium.
(a) One blade of the scissor is placed into the foramen magnum and a lateral cut made. This cut is extended forward as far as the rostral pole of the brain. An identical cut is then made on the opposite side. The blade of the scissor is kept parallel to the brain's surface to avoid cutting it. * = bregma.
(b) The top of the skull is removed.
(c) The rostral tip of the brain is gently lifted upwards, and the optic nerves are severed (*). The brain is then gently pulled backward and out of the cranial vault

the brain is removed. (c) Small or large brain segments break off where the cranial nerves exist. For example, in stretching the optic nerves part of the medial-basal hypothalamus is torn off.

The best practice is to sever the skin from the top of the head with a couple of incisions, and to cut off the occipital muscles from the occipital bone, thus revealing the foramen magnum. One blade of the scissors is cautiously introduced into the foramen. Keeping the blade parallel to the inner wall of the calvarium and as close to it as possible to avoid damaging the brain, cuts are made through the bone (Figure 1a). It is very important that these cuts should extend rostrally as far as the bregma suture, otherwise the bone will break along the suture, and it will be difficult to expose the whole frontal

Figure 2 The brains are mounted on specimen holders. If rostral structures are to be dissected, they are placed forebrain up; if caudal structures are to be dissected, hindbrain up

cortex. After two parallel cuts through the bone are made, one on each side of the skull, the bone is lifted forward and its point of attachment rostrally is broken (Figure 1b). The head is held nose upward and the closed scissors are slipped under the frontal lobe. The brain is gently pried out of the skull (Figure 1c). The brain's movement will be inhibited, however, by the optic nerves which tether its base. Therefore, these nerves are severed to avoid injury to the base of the brain. Finally, the brain is pried out further, the trigeminal nerves are cut, and the brain stem is coaxed out of the skull.

The brain should be balanced on the finger tips as it is being taken from the cranial vault (Figure 1c). Every effort should be made to avoid deforming the brain or causing damage to it which only becomes evident after the sections are made. It helps to wet one's scissors and fingers with saline just before taking the brain out.

If the fresh brain is to be sliced, it should next be placed on a cold plate or in the slicing template where the cutting will be performed. If frozen sections are to be made, the brain should be frozen on to the specimen holder (chuck) of the microtome. Depending on the brain areas that will be dissected, the specimen should be frozen in the 'nose-down' or 'nose-up' position (Figure 2). The frontal cortex (if the brain is nose-down) or the medulla-cerebellum area (if the brain is nose-up) should be cut in the coronal plane, to provide a flat, well-oriented surface to put on the chuck. These cuts should be at least 2 mm to 3 mm from the area that is going to be dissected. It is advantageous to mount the shortest possible brain segment because long brain segments are prone to break off the holder during sectioning. To fix the brain to the chuck saline or Tissue-Tek® (Lab-Tek Products, Naperville, Illinois) may be used. The brain surface that is applied to the chuck should be as large as possible, because cutting thick sections causes considerably more force to be applied to the specimen than cutting thin sections does.

If the specimen cannot be sectioned immediately, it can be stored. It is possible to place the brain into liquid nitrogen immediately after it is removed. This often results in the brain's fracturing, however, and sectioning is then difficult. This problem can be avoided to some extent by allowing the brain to warm up slowly. Remove it from the liquid nitrogen and place it in a cryostat until it is soft enough to cut, remembering that it takes hours before the temperature of the specimen reaches that of the cryostat.

As an alternative to storing the frozen brains in liquid nitrogen, consider placing them in a freezer (−60 °C). They can be stored attached to chucks or the specimen can be placed on a piece of aluminum foil, covered with powdered dry ice, and stored unmounted after freezing. The brain surface that is going to be fixed to the chuck should be applied to the aluminum foil, otherwise the brain may be deformed. The frozen brains should be wrapped in an airtight package (a sealed plastic bag, for example) because otherwise they will desiccate and become impossible to section and dissect. To attach a frozen brain to a specimen holder, allow the part of the brain in contact with the holder to melt slightly or it will not bind well.

The brains of animals killed by microwave irradiation present a special problem. They are difficult to remove from the cranial vault because the meninges fuse the brain to the inner surface of the cranial bone. Furthermore these brains are quite friable. The cranial bones both dorsal and ventral to the brain should be removed in small bits with a rongeur as completely as possible. Try to free the brain completely from adherent structures surrounding it and then remove it. It is sometimes helpful to embed the brain in Tissue-Tek® before sectioning it rather than just attaching it to the specimen holder by its base.

3. SECTIONING THE BRAIN

Microdissection of a sliced brain with either a knife or punch is easy to do. It is possible, with practice, to remove a number of neighbouring structures from the same specimen. Sometimes dozens of specimens are taken from a single brain slice. Furthermore, it is easy to orient the sections in the frontal plane and there are several maps available for the identification of specific cell-groups (König and Klippel, 1963; Hurt *et al.*, 1971; Jacobowitz and Palkovits, 1974; Palkovits and Jacobowitz, 1974; Palkovits, 1980).

The thickness of the section will depend on the technique chosen for slicing the brain and the areas to be removed. Fresh brain slices should be as thin as possible, around 0.5 mm. Extremely thin frozen sections can be made, but in practice 300 μm sections have proven most useful. When thinner sections are used it is difficult to remove the newly dissected tissue samples from the lumen of the punch (see below). Moreover, when thinner sections are prepared more of them have to be cut and dissected without any resulting advantage. Sections thicker than 300 μm (up to 1 mm or 2 mm) can be cut, but many brain nuclei do not extend this far in the anterior–posterior plane. The final decision about the thickness of the sections cut should be made on the basis of the length of the structures to be sampled.

Fresh brain slices may be made manually or with the aid of mechanical devices.

3.1 **Manual brain slicing**

Frontal sections should be prepared. The slicing should be carried out either with a wet razor blade or knife, and after the brain and one's fingers have been dampened with isotonic saline. The brain is placed, basal surface upward, on a black hard rubber plate (Figure 3). The locations of the knife cuts can be determined by placing the blade in front of or behind one or more clearly discernible structures (olfactory tubercle, optic chiasm, mamillary

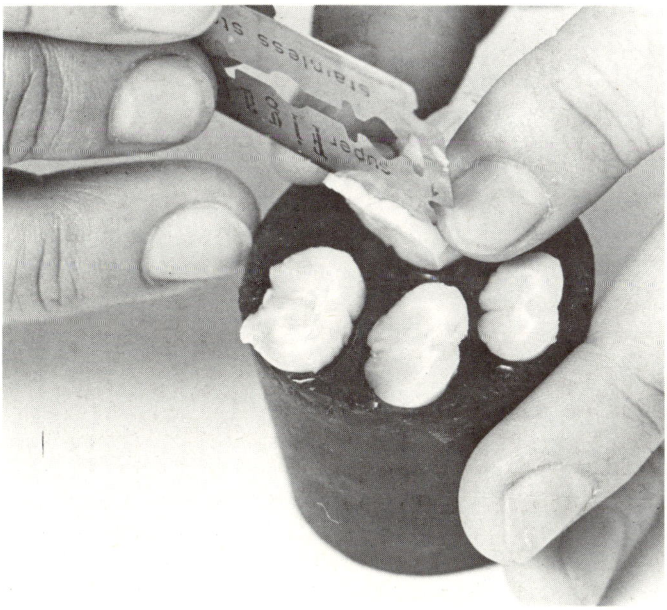

Figure 3 Fresh and fixed brains can be sliced freehand with a stainless steel razor blade. The brains should be placed with their basal surface up on a hard rubber 'stage'. They should be kept moist while they are cut

body, pons, etc.). The brain should be cut through in a single 'pulling' stroke. The brain slice usually remains on the blade, and it should be pushed off gently. With practice 0.5 mm to 1.0 mm slices can be cut. The slicing may be conducted under a stereomicroscope or with the help of a loupe.

3.2 **Brain slicing with devices**

Templates or tissue choppers may be used.

1. A hole is milled in a Plexiglass block corresponding to the shape of the brain. Cuts are made at 1 mm intervals across the block that razor blades can slide into. The brain is placed ventral side up into the block and moistened with saline solution. Then razor blades are inserted into the slots and pressed through the brain (Figure 4). The number of blades used will be determined by the number of slices desired and by the thickness chosen.

Figure 4 The template was fashioned from a block of Plexiglass. A hole the shape of the brain was milled into the block, and then parallel slots were cut perpendicular to the long axis of the brain at 1 mm intervals. The brain is placed into the depression in the block and razor blades are pressed into the slots as shown

2. Tissue choppers have also been used for preparing slices suitable for dissecting. Various types are commercially available: Sorvall TC2 Tissue sectioner, McIlwain tissue chopper (Mickle Laboratory Engineering, Co.), etc. Slices 1.5 mm to 2.0 mm in thickness are produced. These slices can only be used to sample larger rat brain nuclei (caudate, substantia nigra, central grey matter, etc.). In larger species tissue choppers may be used following gross dissections of the brain to prepare the tissue for finer dissection. For use of tissue choppers for microdissection purposes see Chapter 2, Section 2.2.

3.2 **Sectioning frozen brains**

Frozen brains are relatively easy to section, and many chemical constitutents of the brain do not change qualitatively or quantitatively in frozen tissue. The sectioning may be performed either with a freezing microtome or in a cryostat. The latter is highly recommended for speed and simplicity of operation.

A precision cryostat is not required since 300 μm thick sections are to be prepared. However, the instrument should meet the following criteria: the

Figure 5 The specimen holder or chuck can be made of either brass or aluminium. The ball on the bottom of this chuck fits in a socket in the microtome. The chuck can be rotated freely before it is fixed in place

specimen holder should, ideally, be mounted in a ball joint (Figure 5); advancing the specimen holder by 300 μm should be simple to do; and there should be ample space for manipulations within the cryostat with both hands. The following commercial cryostats are suitable for this purpose: Cryo-cut (American Optic), Minotome (IEC).

The ideal temperature for preparing thick sections is about −6 °C to −10 °C. The brain, frozen on to the specimen holder with dry ice, should be placed into the cryostat prior to sectioning, and ample time should be allotted for it to warm up. If the brain gets too warm during sectioning, it can be cooled with powdered dry-ice. Sometimes, it is necessary to cool the knife blade as well. It is occasionally helpful, if the block is too cold, to touch the surface to be sectioned with one's fingers. This warms it somewhat, allowing it to be cut and flattened on to the knife blade more easily.

Sectioning proceeds as follows. The specimen holder is fixed (Figure 6) on to the microtome and positioned, by eye, as close to the correct plane as

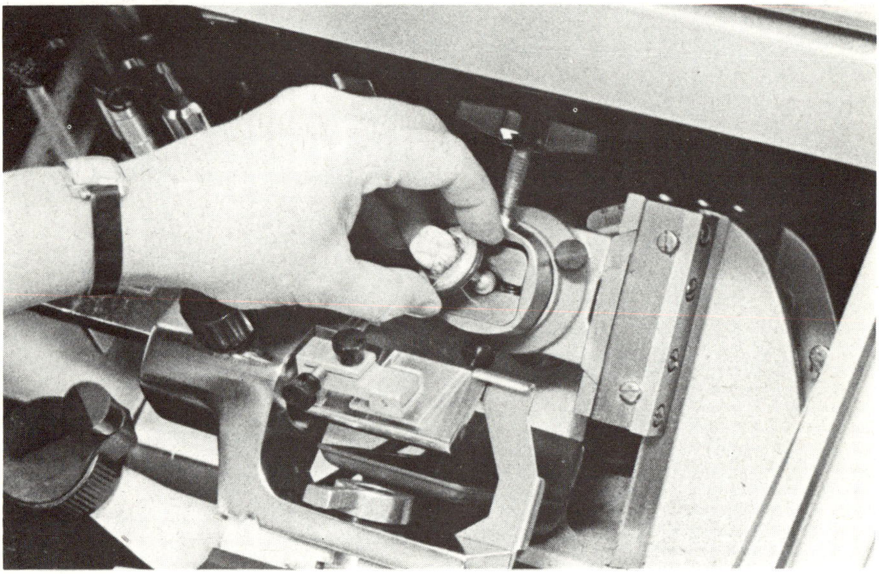

Figure 6 After the brain is mounted on the chuck, the latter is attached to the microtome. The chuck is moved about in order to position the brain properly, and then fixed in place

possible. The advancing mechanism of the microtome is taken out of gear so that the sample can be advanced or withdrawn hundreds of micrometers by turning the advancing mechanism (a metal disc) manually. The disc can be calibrated by observing how much it turns per distance advanced when the advancing mechanism is in gear. The rim of the disc should be marked at 100 μm intervals.

The optimal slope of the knife is 15 °. A plastic plate (anti-roll device) should be positioned above the knife. It should be adjusted in such a way that the 300 μm section travels smoothly across the blade without folding or cracking (Figure 7). The brain should be placed on the specimen holder so that the knife blade does not first contact an area to be sampled, because the structure that first encounters the blade is always damaged slightly (Figure 8). In general, one should begin cutting sections well ahead (on the block) of the region to be dissected. This is necessary to have ample space for correcting the plane of sectioning. The block needs to be adjusted so that it is bilaterally symmetrical (i.e., adjusted to the left or right) and so that it is aligned properly in the dorsal–ventral plane. There are several landmarks in various parts of the brain that will help one orient the brain properly. These have been described earlier (Palkovits, 1980).

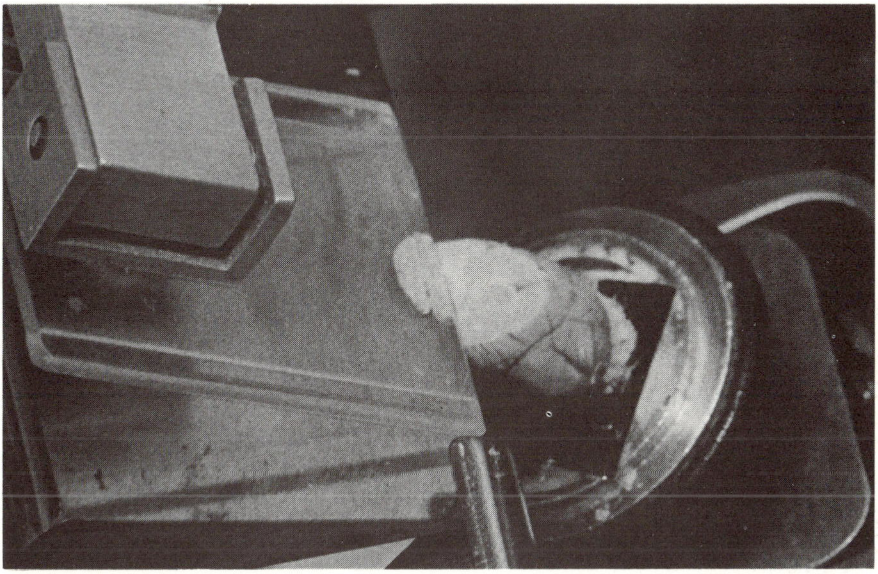

Figure 7 An anti roll device is used to cut flat sections. This plastic plate must be positioned slightly above the blade so that the slice can pass under it

After the specimen has been adjusted, the sections to be dissected can be cut. It is not necessary to place the sections on a glass slide immediately after they are cut; the procedure is more rapid if a few sections are cut (there is place for about three to five sections on the knife) and then transferred to a glass slide.

3.3 **Sectioning irradiated brains**

Free-hand cutting of irradiated brains should not be attempted; irradiated brains should be cut in a cryostat. These brains need to be sectioned at a

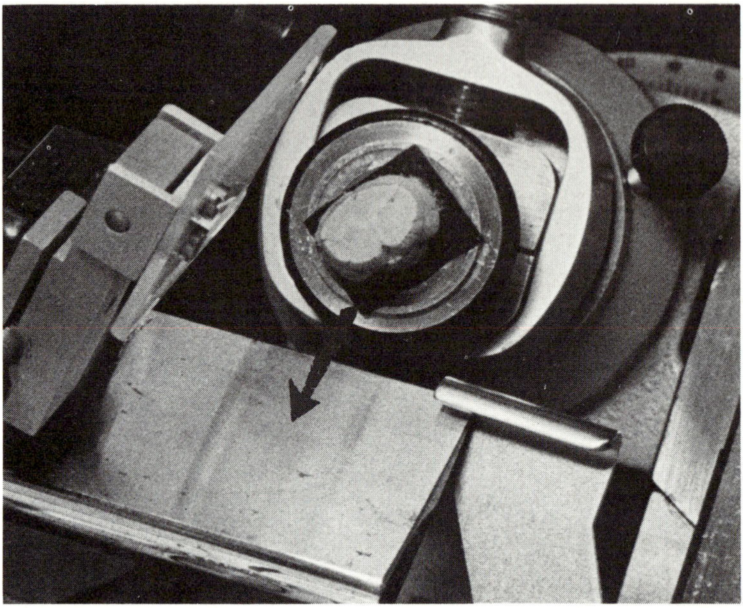

Figure 8 The brain is oriented in such a way that the structures of interest are *not* the first to contact the blade. The medial basal hypothalamus of this brain will be dissected. Therefore, to avoid damaging basal hypothalamic structures (e.g., the median eminence and arcuate nucleus), the brain is oblique to the knife blade

lower temperature (−15 °C) than fresh brains. It is helpful to cool the knife blade intermittently with dry-ice. It is best to section the shortest brain segments possible because the frozen irradiated brain easily breaks while it is being sectioned. As mentioned above, embedding the brain in Tissue-Tek® is useful to provide support for the tissue.

3.4 Sectioning fixed tissue

Microdissection of fixed tissue is mainly useful as an adjunct to electron microscopic studies. Generally, brains fixed by perfusion are easy to section. In fact, they can be cut manually more easily than fresh brains. One must proceed cautiously, as, contrary to the fresh brain, it is easy to break off one or two brain regions while slicing the fixed brain. The brain slices once cut should either be put back into the fixing solution or should be moistened with it; they dry out quickly. They should be kept moist while they are dissected too. It is most convenient to place the fixed brains on a hard black rubber surface while slicing them.

Fresh and fixed brains should be dissected immediately. They should be placed on a black rubber plate under a stereomicroscope. The frozen sections, on the other hand, should be mounted on a glass slide, after which they can either be stored or punched.

These microscope slides should be numbered prior to sectioning the tissue then put into the cryostat, since the frozen sections can only be placed on cool slides (the sections will stick to the warm slides and cannot be repositioned once put down). The sections are transferred to the slides with the help of a brush or forceps. The brush or the forceps have to be stored in the cryostat, otherwise the sections will stick to them. Depending on the size of the brain sections five to ten can be stored on one slide. On a cool slide they may be moved about easily and placed in a row. After this is done the brain slices have to be fixed to the slide. Remove the slide from the cryostat and allow it to sit at room temperature on a desk surface or on the palm of the hand until the sections have just thawed. Then place the slides on a bed of powdered dry-ice. The slightly thawed sections are welded to the glass slide when they refreeze. The entire procedure takes a couple of seconds, so that neither the tissue nor its biochemical constituents are affected in most cases.

After the sections are removed from the knife blade, the top and bottom of the blade have to be wiped off. If any residual brain tissue remains smeared on the knife it should be wiped off with a cloth dampened with alcohol.

The brain slices that have been mounted may be dissected immediately or stored for varying periods of time. When large numbers of animals are studied, sectioning of the brains may require one or more day's work, and storing the sections may be essential. The mounted sections may be placed in slide boxes or, if they are well frozen, simply stacked on one another. They should be placed in an airtight container (e.g., plastic bag) and stored on dry-ice or in a deep freeze.

4. TOOLS FOR PUNCHING

4.1 **Magnifying glasses, loupes, microscopes**

Except for unusually large areas (cortex, caudate, cerebellum, etc.) punching is conducted under a microscope. Any kind of stereo or dissecting microscope is suitable for this purpose. Little (six- to 20-fold) magnification is needed.

A magnifying lens may be used if it magnifies at least five times and if the lens can be placed at least 10 cm above the table on which the dissection will be performed.

A loupe is most convenient to use if the five-fold magnification is provided by a single magnifying lens and not by a compound lens which yields such a narrow field of vision that manipulations with the dissecting needle are difficult to follow.

4.2 **Brain sections must be properly illuminated**

A cool microscope lamp that illuminates the sections from above can be used with a stereomicroscope, magnifying glass, or loupe. Fibre optic systems are especially nice to use because they will not heat up the sections and can be directed easily at the area to be dissected. The sections can be transilluminated if a special cold stage is built which can accommodate the fibre optic light beam. Somewhat more detail can be perceived when the sections are transilluminated, but except in special cases, this is not enough to warrant the extra cost of the cold stage.

4.3 **Sections must be kept frozen while being dissected**

This may be done in one of several ways. A petri dish filled with powdered dry ice makes a fine cold stage. Alternatively the bulb that illuminates the stage of many stereomicroscopes from below may be removed from its housing and the latter filled with dry-ice (Figure 9). Indeed, any type of box

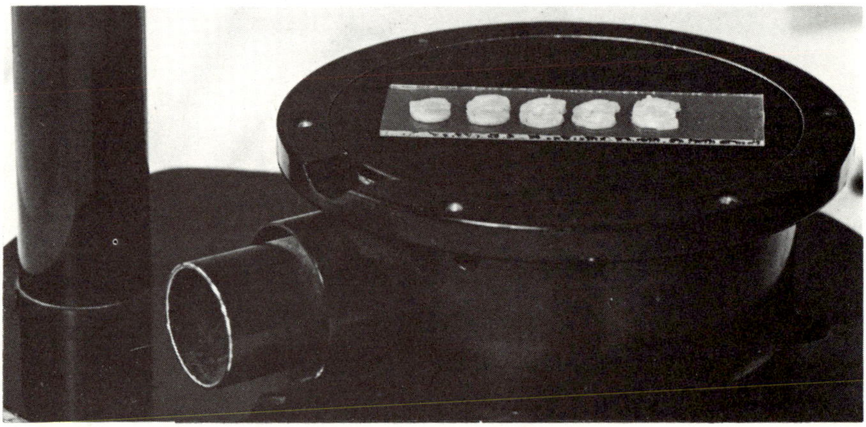

Figure 9 This cold stage has been made by removing the lamp from its housing in the base of a dissecting microscope and replacing the glass top of the lamp box with a black metal plate. Alternatively, a petri dish full of powdered dry-ice can be used

may be used as long as it meets the following requirements. (a) It should ensure constant temperature; (b) its upper lid should be a metal one, or it should be covered with a metal plate that provides good heat conductivity; (c) this cover should be black, if possible, to provide a contrasting background for the sections placed on the glass slide; (d) if a stereomicroscope is used for dissecting the box should not be so high as to prevent focusing.

Specific containers equipped with a circulating cooling system may be used and the need for adding dry-ice to the box from time to time removed. The temperature of a cold stage may be kept fairly constant so that the sections neither get too cold and brittle or too warm and soft to punch. The optimal temperature of the stage is in the range −25 °C to −35 °C.

5. MICRODISSECTION NEEDLES

The hollow punching needles are the most important tools used for the punch technique. Their inner diameters vary from 0.2 mm to 2.0 mm. The size

Figure 10 Microdissection needles.
(a) The round punch is the most frequently used for sampling fresh or frozen tissue.
(b) Square or rectangular punches should be used to dissect lamina structures.
(c) Punches with indentations are used to prepare samples for electron microscopy. The nick in the circle allows one to keep track of the orientation of the sample

chosen is dictated by the cross-sectional diameter of brain region to be dissected.

The length of the needles is 4 cm to 6 cm (2 in to 3 inches). Shorter needles are uncomfortable to hold, longer ones are too bulky to manipulate beneath the stereomicroscope. It is important that the bevelled cutting edges should always be sharp. Punching is conducted on a glass slide at relatively high pressures; after a couple of thousand of punches the needles have either to be sharpened or replaced. Consequently the needles should be constructed from hard stainless steel tubing fixed in a thicker handle (Figure 10). The handle should not be thicker than a pencil, otherwise it will obstruct the view of the needle tip under the microscope (punching is performed with the needle nearly vertical). The handle should not be too thin either because it will hurt one's finger after a while. There are various needles in use:

1. Punching needles with a circular lumen (Figure 10a). These are used most frequently. Needles with inner diameters of 500 μm to 1000 μm (and rarely 300 μm) are most useful for microdissecting rat brains.
2. Specific punching needles with non-circular lumens (Figure 10b). It is more difficult to construct needles with oval or square tips, but these can

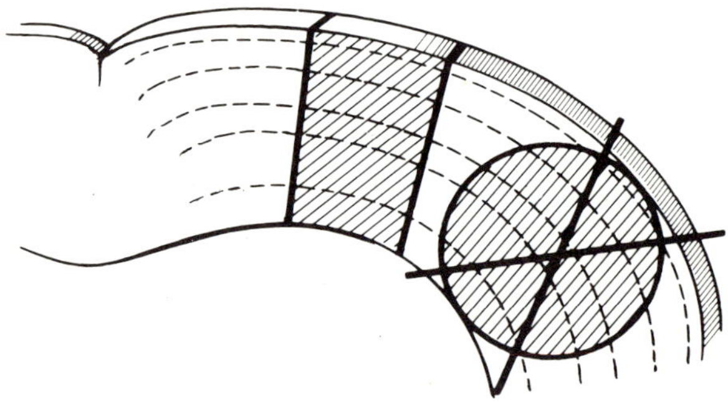

Figure 11 Rectangular punches are used to dissect laminar structures such as the cerebral cortex, olfactory bulb, olfactory tubercle, cerebellum, hippocampus, and superior colliculus. Each layer contributes to the sample in proportion to its individual thickness. Some regions (e.g., cerebral cortex) can be subdivided into superficial, middle, and deep layers with rectangular punches

sometimes be used to dissect brain regions that are hard to remove with round punches. Square tipped needles are used for the sampling of layered brain areas (cerebral cortex, cerebellar cortex, olfactory tubercle, hippocampus, dentate gyrus, superior colliculus, etc.) (Figure 11). If no square needle is available, it is better to sample laminar areas by means of a fine knife than to use a circular needle.
3. Punching needles equipped with a marker in the lumen (Figure 10c) are mainly used for removing tissue prior to electron microscopic analysis. A fine steel thread or needle is fixed into the lumen of the punch, or the needle wall is pressed sharply into the lumen on one side. Using this type of punch requires practice; the hand must keep very still during the dissecting process. In the course of punching the needle is oriented over the section so that the 'mark' always appears in the same place (medial, dorsal, etc.). The nick on the tissue disc is clearly visible, and it can be used to align the section after it is embedded and sectioned.

4. Microdissection needles equipped with a stylet. The punched specimen is usually blown out of the microdissection needle. Sometimes this is undesirable (e.g., when collecting fresh brain specimens for tissue culture) or not preferred by the individual doing the dissection. Then the tissue discs may be pushed out of the needle with the air of a stylet (Figure 12). The stylet should fit fairly tightly into the lumen of the needle or the brain

Figure 12 A dissecting needle and stylet. The stylet is kept from slipping all the way through the punch by a broad shoulder (arrow) that is too big to pass into it. When depressed, the stylet can extend at most 2 mm to 3 mm out of the tip of the dissecting needle (see Figure 13)

tissue will be smeared on to its wall. The stylet should reach no farther than 2 mm to 3 mm beyond the needle tip when it is fully depressed (i.e., there should be some mechanism which prevents the stylet from travelling straight through and out of the punch).

The use of dissecting needles with stylets takes practice. The stylet should be partially withdrawn end pressed against the side of the needle with one of the fingers (Figure 13a) while holding the needle with the others. The needle is pressed into the tissue which should cut easily (like hard cheese) (Figure 13a). The stylet should be released only after the punching is concluded, and

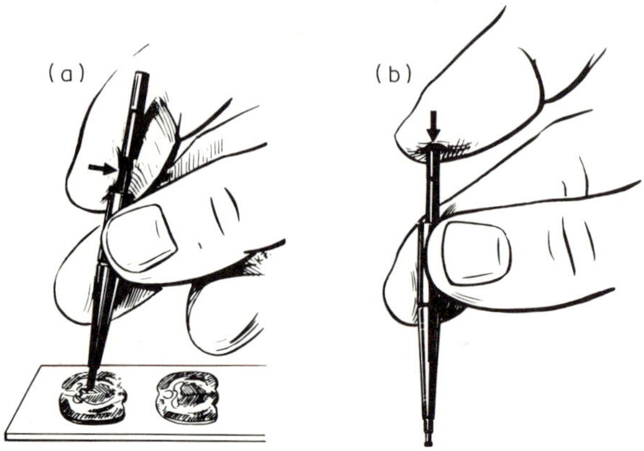

Figure 13 Punching and recovoring tissue samples.
(a) The punch is held between the thumb and middle finger and the stylet is held up with the index finger.
(b) After the samples are punched from the sections, the stylet is released and the samples are pushed out of the dissecting needle

the tissue specimen is ready to be expelled from the needle (Figure 13b). If the stylet is close-fitting, and is equipped with a suitably heavy 'head', releasing it results in its falling down, and it will push the tissue disc out of the needle.

6. THE PUNCH TECHNIQUE

6.1 **Practical advice**

6.1.1 *Training and practice*

Mastery of the punch technique can only be acquired by practice. There are four things which have a major impact on one's ability to do microdissections and which need to be learned at the bench.

1. The moving of one's hands, and the slice, the following of the course of the needle under the microscope, outside the vision of the microscope and repeatedly under the microscope, must all be practiced. These gestures have to be conducted rapidly and automatically. However, this practice can readily be mastered, and one will adapt to working under the microscope.

   The slide (or the rubber plate in the case of fresh brain sections) is moved with one hand, while the needle is held, and the punching performed with the other. The section may have to be moved in either direction (Figure 14). After moving the section into position under the microscope, the slide (or rubber plate) has to be held with two fingers (the hand must not occupy the light path). The needle is held by the first and third fingers of the other hand, while the forefinger is used to adjust the slope of the needle (Figure 14). It is helpful to steady this hand on the microscope stage so it is advisable to protect the fifth finger with a glove.

Figure 14 The slides are positioned with one hand and the punch is held in the other. The slides are moved up and down and from side to side to bring structures to be dissected into view

2. Choosing the proper illumination and magnification. The section ought to be illuminated by a well focused, strong light—one that does not warm it up, however. Since both hands are used under the microscope, the light should be directed in such a way that one's hands do not obstruct it. Ideally a microscope with epi-illumination should be used, but these are quite expensive. The sections must be magnified so that landmarks needed for microdissection can be identified. Too much magnification is not advantageous because it fails to provide a clear survey of the entire section. Each individual needs to choose the most convenient magnification to use based on the material to be dissected, the landmarks that have to be visualized, and the quality of the sections. It is most impractical continually to alter the magnification in the course of microdissection.

3. In the case of fresh or fixed brains the temperature of the sections does not affect the dissection, but in the case of frozen sections it is of major importance. It is difficult to describe the optimal temperature for punching, this has to be set empirically. Sections that are too cold are prone to break while sections that are too warm melt beneath the needle so that sharp-edged circular samples cannot be punched and taken up into it. Thus the temperature of the dissecting surface has to be controlled by the person who is doing the punching unless an automated system is available. The container, when it is freshly filled with dry-ice, will be cooler than

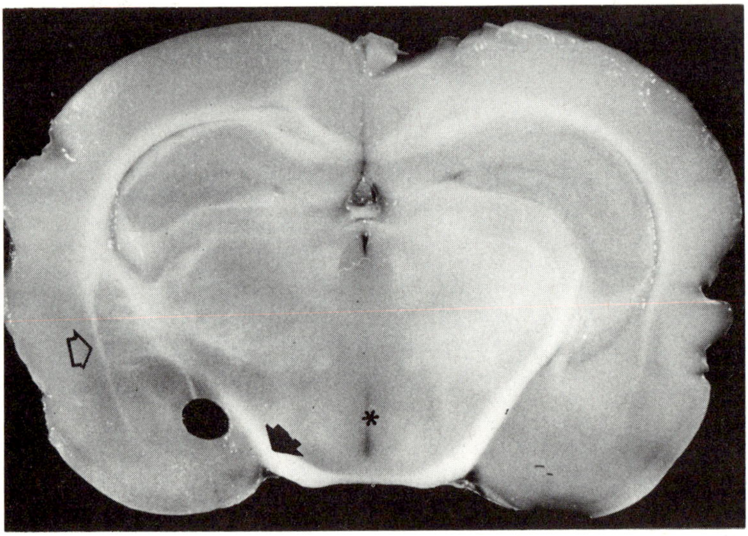

Figure 15 On fresh sections both thick and fine myelinated fibre systems can be visualized. The ventricles can be seen as well. The central nucleus of the amygdala has been removed from this section. (Solid arrow, optic tract; hollow arrow, external capsule; asterisk, third ventricle)

necessary. Therefore, at first, the sections have to be warmed—by waving them in the air or placing them on the palm briefly—to be punched. In time, the dry-ice will disappear from the container, and the top (stage) will warm up slightly. Then the sections will not need to be pre-warmed. Subsequently the stage will become too warm. At that point the addition of a small amount of dry-ice to the box will restore the optimal temperature. With practice temperature control becomes second nature and will not require much conscious effort.

Usually there are several sections per slide. Since several specimens are removed from each section, the slides often have to be under the microscope for a matter of minutes. Under these circumstances even with the best cold stage, the sections need intermittently to be placed on dry-ice. Therefore, it is helpful to keep a container filled with dry-ice powder or with flat pieces of dry-ice, near the microscope (this may also be used for cooling the punches, forceps, microtome knife used for sectioning and dissecting).

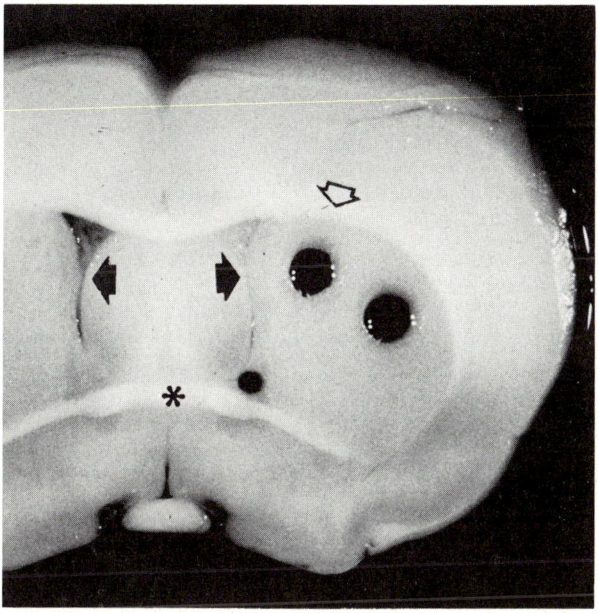

Figure 16 Formalin or glutaraldehyde fixed brains are easy to slice. Thick fibre bundles and ventricles can be seen on these sections. (Solid arrows, lateral ventricles; hollow arrow, corpus callosum; asterisk, anterior commissure.) Two large (1 mm diameter) samples have been removed from the caudate nucleus. One smaller (0.5 mm diameter) sample of the interstitial nucleus of the stria terminalis has been punched as well

Sampling should be performed as quickly as possible because frost accumulates on the surface of the frozen section. The frost can be wiped off with a cloth when it begins to obstruct vision.

4. A dissection protocol should be prepared before the animals are killed. The brains should be sectioned bearing in mind the samples to be removed. Sections should only be prepared from those brain regions which will be dissected. The samples should be removed so that the dissection proceeds in an orderly way from rostral to caudal sections or *vice versa*. The proper needle has to be chosen in advance of punching. Both the size and shape of the structures to be removed must be taken into account. The

number of tissue discs to be removed from each brain should be considered too, because it is helpful for the amount of tissue in each homogenate to be about the same.

A sufficient number of homogenizers or microfuge tubes and of test tubes should be prepared and numbered. The volume of the liquid required for homogenization depends on the size of the tissue samples. A minimum volume of about 20 μl is required if microhomogenizers are used. The solution in which the samples are homogenized depends on the assay or assays to be performed.

6.2 **Topographical orientation; landmarks**

The white matter can be seen on fresh sections of the brain as long as the sections are not allowed to dry (Figure 15). These sections are easy to orient provided that they are placed on a black background and that there is good illumination available.

Orientation is much improved if the brain tissue can be stained prior to microdissection. Zigmond and Ben-Ari (1976) have applied methylene blue

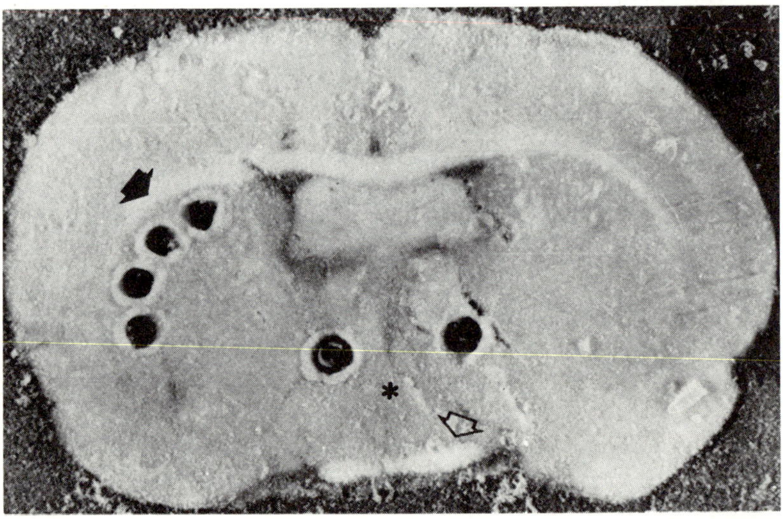

Figure 17 This frozen brain section has accumulated a modest amount of frost. In spite of this the ventricles can be visualized, as can major white matter structures. Four samples of caudate nucleus have been removed from the inner edge of the corpus callosum (solid arrow) with 0.7 mm diameter punches, and two samples of the caudal part of the interstitial nucleus of the stria terminalis have been dissected, one from each side of the brain. The hollow arrow points to the optic chiasm; the asterisk is beside the third ventricle, which can be seen more clearly under the microscope than in the photograph

stain to sections. This enables them to visualize groups of cells under the microscope.

In fixed sections fine fibres cannot be visualized, only large fibre bundles. The ventricular system, however, can easily be recognized (Figure 16). Since fixed tissues will not undergo changes *post mortem*, there is ample time to study them under the microscope and change the magnification that is being used if necessary.

The white matter and the ventricles can be perceived easily on frozen sections (Figure 17). Large brain areas can easily be identified on these

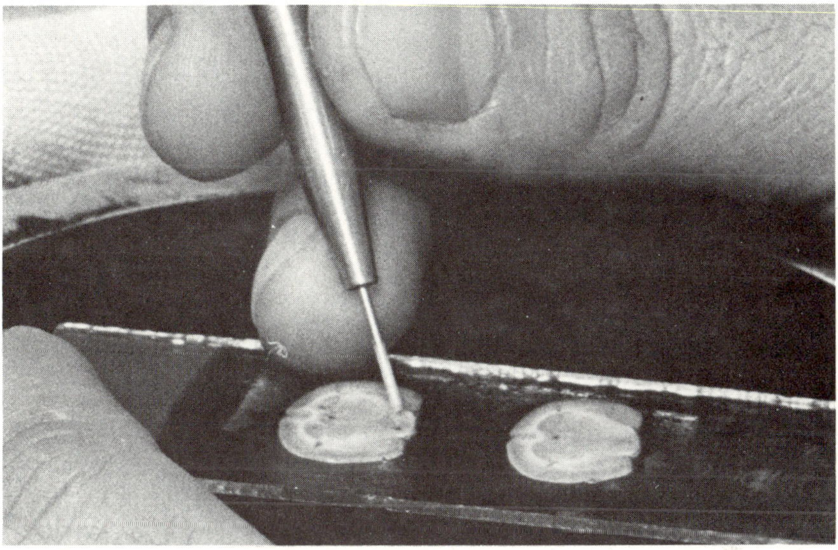

Figure 18 The punch is held at an oblique angle and its tip is positioned over the structure to be dissected

sections. Difficulties only arise if the sections are not in the coronal plane, if they are asymmetrical, or if the brain is damaged as a result of experimental intervention or sloppy removal. Under these circumstances the investigator must decide whether the landmarks that can be recognized will guarantee accurate sampling. When in doubt, the dissected sections should be saved for histological examination.

6.3 **Micropunch**

To position the tip of the punch, the section is approached with the punch held at about 45° to the surface of the section (the handle of the needle should not impede visibility). When the tip of the needle is in the appropriate location (Figure 18) it should be fixed with gentle pressure to the surface of

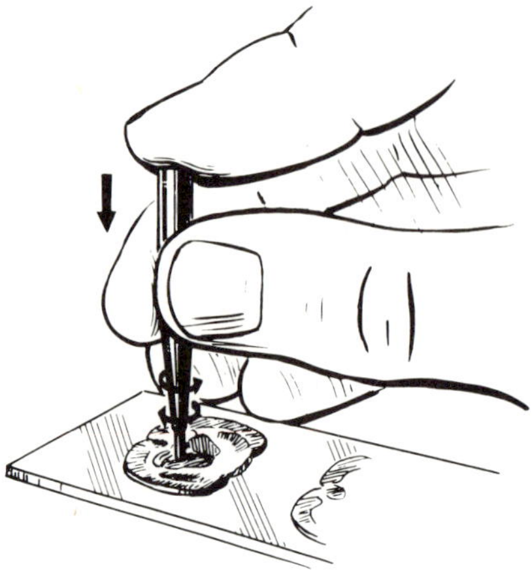

Figure 19 After the tip of the needle is in place it is pressed against the section and anchored there. Then the punch is brought into a vertical position and pressed into the brain slice. Finally the punch is rotated to free the sample from the surrounding tissue, and withdrawn

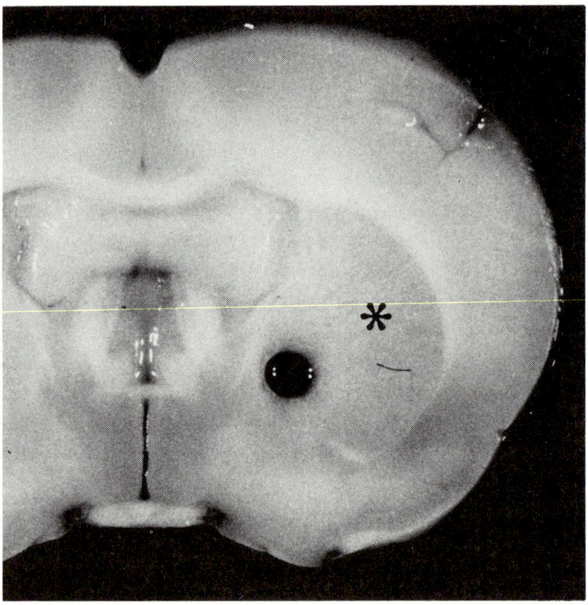

Figure 20 The margins of the hole left after this disc of globus pallidus was removed are nice and sharp. No tissue fragments were left behind. The asterisk indicates the adjacent striatum

the section and the punch should be brought into a vertical position. Then the needle can be pressed into the tissue, rotated slightly, and abruptly withdrawn (Figure 19).

After a tissue sample is successfully withdrawn a sharp edged hole remains in the section (Figure 20), and the tissue disc is stuck in the needle. If tissue segments are retained in the punched area, or if the entire sample remains behind, the needle is replaced in its original position and the punching is repeated.

The tissue disc is either blown out, or, if a needle with a stylet is used, pushed out of the punch into the bottom of the homogenizer or tube. Sampling of a discrete brain area may be performed by punching several sections in succession. The discs need not be blown out separately; six to eight tissue discs may be collected in the lumen of the needle and removed all at once.

If several brain areas are to be punched from a single section, the first areas removed should be those nearest the edges of the section. Occasionally (especially in cold sections) the section may get cracked or even broken during the punching procedure. The sections typically crack in the direction of the closest surface. If a section cracks, raise its temperature until it starts to soften and then attempt to continue.

To remove a very small area (with a 0.2 mm to 0.3 mm needle) it is best to cool the needle tip on dry-ice for a couple of seconds beforehand. This facilitates blowing the brain tissue out of the thin needle. The same procedure may be applied if the entire tissue disc is needed to avoid thawing of the tissue within the needle lumen prior to blowing it out.

The dissecting needles need to be cleaned after each experiment. They should be washed thoroughly with saline then with alcohol. Microdissection may also be carried out with a fine knife (so-called Graefe knife) (Figure 21). The procedure is exactly the same as that described above, except that the required brain area is excised instead of punched. The sections can be cut more easily with a knife if they are somewhat warmer than they would be for punching. Using the knife has its drawbacks, however. It is much slower than punching, and less precise. The knife may be used to dissect elongated areas (for instance the periventricular nuclei) or layered regions when proportional sampling is required, and square needles are not available.

7. GENERAL RULES FOR MICROPUNCH DISSECTIONS

Compared to macrodissections the micropunch technique is quite reproducible and the specimens obtained can be well-characterized. Macrodissections are reproducible enough when parts of a large, homogeneous brain area are excised, but is not very reproducible when applied to heterogeneous brain areas, which are, in fact, more common. It is safe to assume that brain areas removed in different laboratories by means of macrodissections are variable in their make-up despite the fact that they are given similar or identical names.

One of the basic rules of the micropunch is always to use a smaller needle than the brain area to be removed. The tissue disc need not contain the entire nucleus, instead the specimen should be taken from the nucleus (Figure 22). The sample should not contain, if possible, foreign (neighbouring) brain

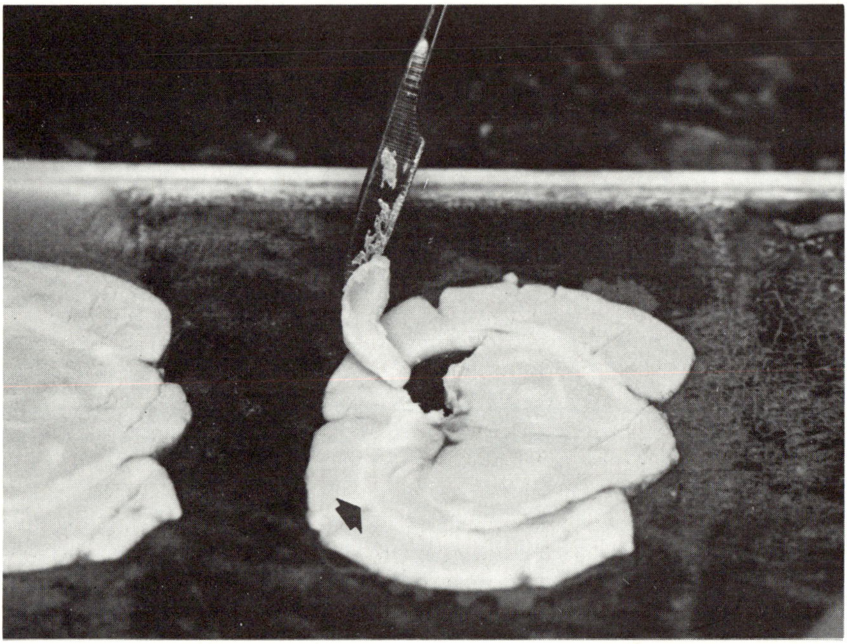

Figure 21 A Graefe knife can be used to dissect large structures from frozen sections. The hippocampus is being removed from this slice. The arrow points to the corpus callosum

areas. However, if this is unavoidable, the contamination should be as small as possible. Ideally, the specimen should contain the entire nucleus, and the amount of 'foreign' material in the sample should be specified.

When dissecting large regions (e.g., the caudate nucleus) the specimens should be taken from the same part of the brain region each time. To the extent that subdivisions of individual nuclei are recognized, they should be considered independent nuclei and separated from one another.

Attempt, whenever possible, to remove complete tissue discs. Unfortunately, the section may be stuck to the slide in such a way that only part of a sample enters the needle. In this event the punching procedure should be repeated. With more pressure and rotation of the needle the tissue fragment can usually be lifted. This is necessary to ensure homogeneous sampling and when it is necessary to estimate the weight of the tissue samples.

There are times (electron microscopic or tissue culture studies) when it is absolutely essential to obtain undamaged tissue discs. With fixed brains these are fairly simple to produce. Fresh brains, on the other hand, must be dissected as soon as possible following sectioning. For complete recovery of tissue samples from frozen sections the punch should be placed on dry ice for a couple of sections prior to punching.

Many tissue discs may be punched from large brain areas. It is useful to limit the sample size though. Having similar samples facilitates homogenization, protein assays, and the standardization of measurements.

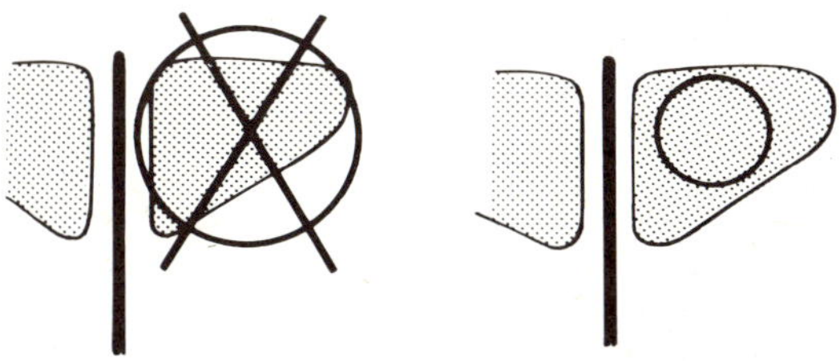

Figure 22 A punch small enough to fit into the structure that is being *sampled* should be used. A punch with too large a diameter has been used to dissect the paraventricular nucleus on the left; the proper sized punch is depicred on the right

The punch technique should be carried out as rapidly as possible. Otherwise the warming of the brain tissue may adversely affect the level of substances to be assayed. Furthermore, the longer a frozen section is kept outside the box of dry-ice used for storage, the poorer its quality. Landmarks get steadily harder to recognize. Fresh and fixed sections gradually desiccate so they should be kept moistened while they are being dissected (isotonic salt solution should be applied to the fresh sections and fixative to the fixed ones).

8. DETERMINATION OF THE SAMPLE SIZE AFTER THE DISSECTION IS COMPLETE

When biochemical measurements are made on tissue extracts, the substance(s) assayed are normalized to the sample size. The sample size may be determined in one of several ways:

If there are milligram amounts of brain tissue, the samples may be weighed. Small samples should not be weighed because they dry rapidly. The ensuing weight loss may amount to between 40% and 60%. If the fresh tissue is to be weighed, the specimens should be stored in buffered saline until the weight is measured. Excess solvent should be removed with a filter paper immediately prior to weighing.

Frozen samples can also be weighed. Freezing the specimens prevents some post-mortem alterations. When frozen samples are weighed at room temperature in the open, water rapidly condenses on them and increases their apparent weight.

Since the protein content of the brain is fairly stable, the amount of protein in a brain extract is proportional to its tissue content. Protein assays may be carried out with the Folin reagent according to the method of Lowry *et al.* (1951). An aliquot of the tissue homogenate is put aside for protein determination; the remainder is used for other purposes. A number of substances interfere with the Lowry assay (e.g., Tris, HEPES) and appropriate reagent blanks, must be run along with the samples.

Before an aliquot is removed for protein determination, the sample must be completely and thoroughly homogenized. No tissue fragments must remain. The homogenate should be vortexed just before the aliquot is removed. The crude homogenate should be sampled, not a supernatant or pellet. A well-suspended crude homogenate will contain both soluble and insoluble tissue components; a supernatant or pellet will contain one or the other.

The amount of the brain tissue in a homogenate can be estimated from its protein content by multiplying the latter by ten; about one-tenth of the wet weight of the brain is protein. The protein content of specific brain areas may deviate from this value somewhat. The protein levels of discrete cell groups or areas of the rat brain have not been mapped as far as we know.

Before deciding to normalize biochemical values to protein levels, the investigator should consider the possibility that the protein content of brain tissue may vary as a result of experimental interventions. If the brain is damaged the change in protein may be considered.

Estimating tissue weight from the size and number of tissue pellets: the volume of the tissue disc may be calculated from the thickness of the section and the inner diameter of the needle. The accuracy of this value is influenced by variations of the width of the section and by the completeness of the sampling. Small tissue fragments may be retained both on the slide and on the inner wall of the needle. If the sections are 300 μm thick and if the discs are perfectly intact, the wet weight of the punched samples (assuming that the density of the brain is 1 mg μl^{-1}) would be the following: with a needle 0.2 mm, 10 μg; 0.3 mm, 20 μg; 0.5 mm, 60 μg; 0.7 mm, 115 μg; 1 mm, 235 μg, 1.5 mm, 0.5 mg; and 2 mm, 1 mg.

9. HOMOGENIZATION OF MICRODISSECTED BRAIN TISSUE

The punched sample is either blown out of the needle or pushed out with a stylet. If the tissue disc is going to be used for histological, tissue culture or electron microscopic studies it must remain intact at the end of the dissection procedure. If biochemical analyses will be undertaken, the tissue sample must be homogenized. Typically, we use ten or more times as much solvent as tissue (vol/vol), but the appropriate volume must be established empirically and based on efficiency of extraction of the substance to be measured. Homogenization may be carried out in one of several ways:

Microhomogenizers: small glass homogenizers are available for manual homogenization in solvent volumes of 20 μl to 200 μl. We have used homogenizers made by the Micro-Metric Instrument Co. or by the Kontes Co., but any small homogenizer will do as long as the mortar and glass chamber are close fitting. After homogenizing no tissue fragments should be detectable in the homogenate. In using these homogenizers the tissue disc is usually blown from the needle onto the end of the mortar.

With high speed homogenizers the glass or metal rotor is driven by a motor and the glass chamber is moved up and down. The tissue disc is blown on to the inner wall of the glass chamber or pushed directly into the homogenizing solvent by the stylet. These homogenizers are simple and fast to use.

Homogenization by sonication. There are diverse ultrasonicators commercially available. The homogenization of tissue discs require about 30 seconds. The time required for homogenizing different brain areas may vary somewhat. In regions rich in fibres (especially in the myelinated fibres) homogenization proceeds rather slowly. Consequently in large scale experiments a multi-tube sonicator is very useful.

10. VALIDATING THE MICRODISSECTION METHOD

The scheduling and frequency of validation of the dissections may be set according to the expertise of the experimentor and the difficulty of the dissection. There are several ways to check the accuracy of the dissections.

After dissecting fresh brain slices (Figure 23) or frozen sections the (thawed) sections can be examined under the microscope. Even small fibres are easily discernible under the stereomicroscope, illuminated from above and placed on a black background. The sections should *not* be allowed to dry. A 20- to 30-fold magnification will permit examination and specification of the site from where the tissue disc was removed.

Both punching and subsequent examination of the slices may be facilitated by staining the sections with a dye as long as this does not interfere with ensuing studies (Zigmond and Ben-Ari, 1976). Stained sections are really no better to study under the microscope than fresh unstained ones (see Chapter 2). The latter are so easy to examine microscopically that no improvement is necessary. Fresh sections do desiccate rapidly though, and then they cannot be studied any longer.

If the fresh section becomes desiccated, it has to be fixed and stained to be preserved and examined. The sections are usually fixed in 4% formalin, dried, stained with hematoxylin-eosin, and covered. The staining reactions should be much shorter than when using thin sections or else the sections will

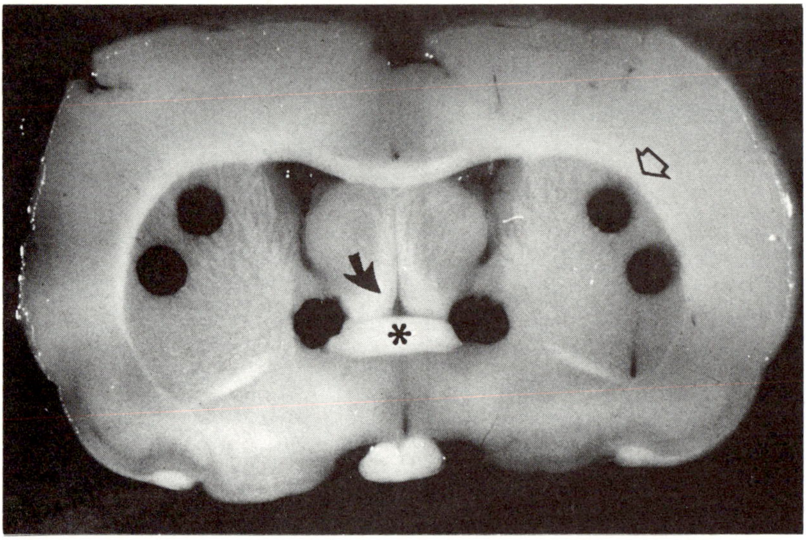

Figure 23 After this 300 μm frozen section was dissected it was allowed to thaw. The white matter can now be distinguished from grey matter readily, and the ventricles are also easy to discern. Two samples of caudate nucleus have been removed from each side of the brain. These were bound dorsolaterally by the corpus callosum (hollow arrow). A sample of the interstitial nucleus of the stria terminalis has been removed from each side of the brain. Medial to these samples is the anterior commissure (asterisk), and dorsomedial to them the fornices (solid arrow)

be too dark. Frozen sections should be fixed immediately after punching, otherwise they are likely to peel off the slide. This is done by immersing the still frozen sections into the fixative.

In many cases insufficient detail can be made out on 300 μm sections. Then thinner sections have to be prepared, either by alternate sectioning, or post-sectioning.

1. In alternate sectioning after each 300 μm section is cut a 10 μm section is prepared. The thin sections are stained and examined for their orientation, effects of surgery, location of micro-injections, etc. Punching is carried out after the thin sections are studied. Though the precision of the dissection

can be improved by this method, the validation of the dissection is still indirect: punched thick sections are compared to the thin ones. Moreover, alternate sectioning is slow, tedious, and requires quite a lot of technical finesse.

2. After the punching is completed, thin sections may be prepared from frozen or manually sliced thick sections. Various embedding techniques

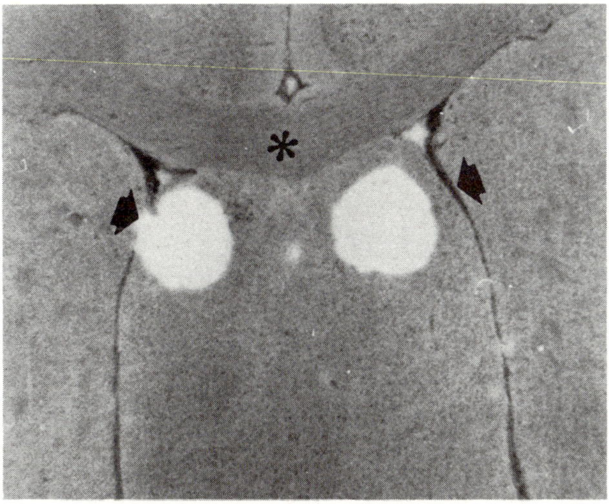

Figure 24 Thick frozen sections can be remounted on microtome chucks after samples have been dissected from them. Thin (20 μm) sections can then be cut, mounted on glass slides, fixed, stained, and covered. These thin sections are not of extraordinarily high quality, but can certainly be used to validate the microdissection method. Here, two samples of dorsolateral septum (solid arrows) have been removed from a thick slice which has then be re-sectioned. The thin sections were stained with haematoxylin-eosin. The asterisk is in the corpus callosum

may be employed, but the thick sections must be handled extremely carefully because they are prone to shrink and curl. Consequently it is nearly impossible to prepare unbroken thin sections in the same plane as the original sections.

It is much simpler to prepare frozen sections. The punched 300 μm brain slices are mounted on a block of gelatin or Tissue-Tek® which in turn is applied to the specimen holder. The block must be 5 mm to 10 mm high and have a flat upper surface. This is easy to form. The block is frozen to the specimen holder, pared down with the knife until it is flat, and only then are the brain sections frozen to it. Subsequently sections of 10 μm to 20 μm are cut, fixed, stained, covered and the dissection validated under the microscope (Figure 24).

11. DISSECTIONS BY ONE OR MORE PEOPLE

Over all, the punch technique is fairly simple. One person can master and perform all the steps, but when large numbers of samples are to be processed it is better to have more than a single individual involved. Therefore, we have described below how to go about the microdissection when one and more than one person is undertaking it.

11.1 **One person**

When samples are being prepared for histology, electron microscopy, or tissue culture, when a small number of samples are being dissected for biochemical analysis, or when the biochemical analysis envisioned does not have to be begun immediately after the dissection is finished, the dissection may be carried out by a single person.

When fresh brains are dissected, the number of brain areas sampled should be limited to ensure that the procedure can be finished in two to three minutes. At longer times the sections get desiccated or gummy due to alterations *post mortem*. In dissecting fixed brains only desiccation has to be avoided. In working with frozen sections the number of experimental animals is limited by the time required for decapitation and freezing the brains to the block. Decapitation, brain removal and freezing cannot be carried out by a single person in less than five minutes; 12 animals may be processed in one hour. The sections are stored in a box filled with dry ice powder, and punching performed after all brains (or a larger number of them) have been sliced. If the experiment is conducted by one person it is advisable to homogenize by sonication, because in this manner a large number of specimens may be dissected before one must stop to do the homogenization.

11.2 **Two people**

If the specimens are removed from fresh or fixed brains, one person should do the perfusions, fixation, decapitation and brain slicing. During the dissection, this person should pipette the homogenizing solvent, number the samples, and help homogenize them. The other person is responsible for the microdissections.

If frozen sections are going to be dissected, one person decapitates the animals and removes the brains, while the other freezes it to the specimen holder. Subsequently, while one person is cutting sections (and storing them on dry-ice), the second person should prepare for the microdissection (number tubes, pipette solutions, prepare the protocol, etc.). Finally, one person does the punching while the other homogenizes. In case microhomogenizers are in short supply, slice only as many brains as can be dissected.

Then, while one person transfers samples to test tubes and cleans the microhomogenizers, the other slices more brains.

11.3 **Team work**

No more than two people are needed to dissect fresh or fixed brains. When punching frozen sections, there is easily enough work for three or four people.

At the beginning of the experiment one person decapitates animals, one removes the brains, and one freezes them to the specimen holders. Subsequently one person slices the brains, one punches, one homogenizes, and one centrifuges the samples after removing aliquots for protein determination.

If punching is not carried out simultaneously with slicing, everything proceeds as above. On the day of the dissection, it is helpful if one person is responsible for handing the slides, in order, to the person who is doing the punching.

11.4 **Optimal protocol for the micropunch dissection**

The micropunch technique enables experiments to be designed that involve assaying hundreds of brain samples. This requires that meticulous and detailed protocols be designed. If there are sufficient people, the following experimental design schedule is recommended:

The number of animals to be used should be determined.

The brain regions to be removed have to be chosen. The number of samples can then be estimated. A dissection protocol should then be prepared—one that takes into account the size of the punch used, the number of pellets removed per nucleus, the number of sections that will be required, etc. In general practice the number of brains to be sectioned and the biochemical procedures that will be employed constrain the experimental design, not the microdissection itself. Under optimal conditions an average of 150 to 200 or more micropunches can be removed per hour.

Twice as many tubes as samples must be made ready: one for the material to be measured and one for protein determination.

The homogenizing solution is pipetted into the homogenizing vessels.

A schedule is set up for mechanical homogenization (or sonication) and centrifugation. (If microhomogenizers are used the time required for washing and drying has to be taken into account; these procedures are very time consuming.)

The brains are sectioned according to the brain areas to be sampled: the means of freezing the brain on the specimen holder ('nose down' or 'nose up'), the level at which the sectioning begins, the sections saved, or discarded, etc. After the number of sections is estimated, the slices are numbered and pre-cooled in the cryostat.

The methods of killing the animal and of removing the brain must be selected. If two people are involved in killing the animals and removing the brains the entire process should take about one minute per animal.

Depending on the number of animals to be studied, the number of sections to be cut, and the number of samples to be punched, the dissection can consume one day or more than one day. You will need to decide whether to do the dissections immediately after slicing the brains or to store the frozen sections. A technically adept person using an adequate cryostat can cut serial 300 μm sections of the whole rat brain in 25–30 minutes. Dissections of specific brain areas require fewer sections. A slide (five sections) can be prepared in about three minutes. One person can serially section 16 to 20 rat brains per day or prepare 120 to 150 slides (from 50 or fewer animals). (This work can be accomplished in eight hours of cryostat cutting. If this task is divided between two people their combined output will not increase much, but the task becomes more manageable.

The protocol for the microdissection is prepared, and the people involved in doing the experiment are each assigned their individual tasks. An experiment with four groups of six or seven animals from which six to ten brain regions will be removed maybe realistically scheduled for one day.

12. MICRODISSECTION OF THE RAT BRAIN

Microdissection of any brain region of any species may be performed. Good morphological orientation is of major importance. Suitable topographical maps are absolutely essential. Unfortunately there are very few maps intended to assist a person engaged in microdissection available. In theory any set of serial sections may be used as a guide as long as similar brain sections can be cut.

Detailed maps that depict the rat brain have been published (König and Klippel, 1963; Hurt *et al.*, 1971; Jacobowitz and Palkovits, 1974; Palkovits and Jacobowitz, 1974). Beyond these, there are complete microdissection maps (Palkovits, 1980; M. Palkovits and M. Brownstein, in preparation). No detailed guide has yet been published describing the topography of individual groups of nuclei, the landmarks that help localize these nuclei, and the size of the punches that should be used to remove discrete nuclei. However, original papers describing the microdissection of various brain regions are available. These include:

Cerebral cortical areas: Palkovits *et al.* (1979).
Thalamus: Brownstein *et al.* (1975).
Hypothalamus: Palkovits (1975), or Palkovits *et al.* (1974a).
Limbic system: Palkovits *et al.* (1974b).

Lower brain stem (mesencephalon, pons, medulla oblongata), Palkovits *et al.* (1974c).

More than 220 nuclei and areas may be excised from the rat brain (Palkovits, 1980). For beginner-anatomists we suggest trying to learn to dissect no more than ten to 15 nuclei at a time. When these can be recognized and their dissection has been validated histologically, then it is safe to start to study a new group.

13. GENERAL COMMENTS

Since its introduction in 1973, the micropunch technique has been adopted by dozens of laboratories. The popularity of the method paralleled the development of neurochemical micromethods. The punch technique offers an order of magnitude better structural resolution than was therefore available, so that now chemical analyses of brain cell groups can be undertaken. Thus, over the last six years neurobiologists have begun to focus on brain nuclei instead of more heterogeneous brain regions. Morphologists have entered into partnership with neurochemists and neuropharmacologists and the interaction has benefited all those involved.

The micropunch technique itself is rather simple; this is the most important cause of its popularity. In this Chapter we have tried to describe our use of the method and to provide helpful hints on its application. We are sure that the method can be improved and refined. In our opinion the best results will be obtained if each person allows the experiments envisioned to dictate the methods finally chosen.

The relative simplicity of the method makes it seductive and potentially harmful. It is easy to punch samples out of brain slices without regard to validation of the dissection procedure. Thus, large volumes of uninterpretable data can be generated.

The punch technique is only one of many methods available to the neurobiologist. Misleading information obtained through its use is a direct result of poor technical or intellectual preparation by the experimenter. Even when it is well-executed, the microdissection method cannot make up for a lack of originality or logic. The method, and this article, should only be seen as adjuncts to the pursuit of these goals.

REFERENCES

Brownstein, M. J., Kobayashi, R., Palkovits, M. and Saavedra, J. M. (1975) 'Choline acetyltransferase levels in diencephalic nuclei in the rat.' *J. Neurochem.*, **24**, 35–38.

Eik-Nes, K. B. and Brizzee, K. R. (1965) 'Concentration of tritium in brain tissue of dogs given (1,2-$^3H_2$)-cortisol intravenously.' *Biochim. Biophys. Acta*, **37**, 320–333.

Hurt, G. A., Hanaway J. and Netsky, M. G. (1971) 'Stereotaxic atlas of the mesencephalon in the albino rat.' *Confin. Neurol.*, **33,** 93–115.

Jacobowitz, D. M. (1974) 'Removal of discrete fresh regions of the rat brain.' *Brain Res.*, **80,** 111–115.

Jacobowitz, D. M. and Palkovits, M. (1974) 'Topographic atlas of catecholamine and acetylcholinesterase-containing neurons in the rat brain. I. Forebrain (Telencephalon, diencephalon).' *J. Comp. Neurol.*, **157,** 13–28.

König, J. F. R. and Klippel, R. A. (1963) *The rat brain: a stereotaxic atlas of the forebrain and lower parts of the brain stem*, Williams and Wilkins, Baltimore.

Lowry, O. H., Rosebrough, N. Y., Farr, A. L. and Randall, R. J. (1951) 'Protein measurement with the Folin phenol reagent.' *J. Biol. Chem.*, **193,** 265–275.

Palkovits, M. (1973) 'Isolated removal of hypothalamic or other brain nuclei of the rat.' *Brain Res.*, **59,** 449–450.

Palkovits, M. (1975) 'Isolated removal of hypothalamic nuclei for neuroendocrinological and neurochemical studies'. in *Topographical Neuroendocrinology* (Eds W. E. Stumpf and L. D. Grant), Karger, Basel, pp. 72–80.

Palkovits, M. (1980) *Guide and map for the isolated removal of the rat brain areas* (Hungarian text), Academic Press, Budapest.

Palkovits, M. and Jacobowitz, D. M. (1974) 'Topographic atlas of catecholamine and acetylcholinesterase-containing neurons in the rat brain. II. Hindbrain (mesencephalon, rhombencephalon).' *J. Comp. Neurol.*, **157,** 29–42.

Palkovits, M., Brownstein, M., Saavedra, J. M. and Axelrod, J. (1974a) 'Norepinephrine and dopamine content of hypothalamic nuclei of rat.' *Brain Res.*, **77,** 137–149.

Palkovits, M., Saavedra, J. M., Kobayashi, R. M. and Brownstein, M (1974b). 'Choline acetyltransferase content of limbic nuclei of the rat.' *Brain Res.*, **79,** 443–450.

Palkovits, M., Brownstein, M. and Saavedra, J. M. (1974c) 'Serotonin content of the brain stem nuclei in the rat.' *Brain Res.*, **80,** 237–249.

Palkovits, M., Záborszky, L., Brownstein, M. J., Fekete, M. I. K., Herman, J. P. and Kanyicska, B. (1979) 'Distribution of norepinephrine and dopamine in cerebral cortical areas of the rat' *Brain Res. Bull.*, **4,** 593–601.

Schumpf, M., Waser, P., Lichtensteiger, W., Langmann, H. and Schlup, D. (1974) 'Standardized excision of small areas of rat and mouse brain with topographical control.' *Biochem. Pharmacol.*, **23,** 2447–2449.

Zigmond, R. E. and Ben-Ari, Y. (1976) 'A simple method for the serial sectioning of fresh brain and the removal of identifiable nuclei from stained sections for biochemical analysis.' *J. Neurochem.*, **26,** 1285–1287.

Brain Microdissection Techniques
Edited by A. C. Cuello

CHAPTER 2

Microdissection of Fresh Rat Brain Tissue Slices

A. CLAUDIO CUELLO* and SUSAN CARSON†

Neuropharmacology/Neuroanatomy Group, Departments of Pharmacology and Human Anatomy†, University of Oxford, England*

CONTENTS

1. Introduction 37
2. Preparation of fresh brain tissue slices 38
   2.1 Intracardiac perfusion with cold solutions 38
   2.2 Cutting sections from fresh brains 39
      2.2.1 The McIlwain Tissue Chopper 39
      2.2.2 The Oxford Vibratome 45
3. Microdissection of fresh brain slices 46
   3.1 Materials 46
   3.2 Procedure 49
4. Micrographs of fresh rat brain slices 51
   4.1 Description of micrographs 51
   4.2 Identification of microanatomical areas in fresh tissue slices of the rat brain 113
      4.2.1 Abbreviations to Figure 6–35 113
5. Examples of microdissected areas of fresh rat brain 117
6. Applications of the technique 121

1. INTRODUCTION

Parallel to the development of highly sensitive radiochemical assays able to detect putative neurotransmitters, enzymes, and receptors in picomole quantities came the development and increasing use of simple, reproducible microdissection techniques. One of the more well-known methods is the 'punching' of frozen tissue samples developed by Palkovits and collaborators (described in Chapter 1) which allowed the processing of large numbers of samples for neurochemical analysis. In this Chapter we describe the microdissection of fresh brain slices as performed in our laboratory. This technique is based on the original procedure of Zigmond and Ben-Ari (1976). However, instead of staining the tissue with methylene blue we base our identification of nuclei and fibre tracts on the differential transparency of these structures to the passage of light by transillumination of the tissue slices. In this manner myelinated structures appear dark and nuclei transparent.

In the following sections we describe in detail the procedure for microdissection of fresh tissue slices as well as the material needed. This approach can be modified depending upon the needs and facilities of individual laboratories. In addition we include a series of micrographs of the fresh rat brain and some examples of different microdissected areas. Finally, we discuss the advantages and applications of this procedure.

2. PREPARATION OF FRESH BRAIN TISSUE SLICES

2.1 Intracardiac perfusion with cold solutions

Although it is possible to obtain good fresh tissue slices without prior treatment, it is advantageous to perfuse the brain with ice cold Krebs' solution (Krebs' Henseleit containing ascorbic acid (130 μM) and ethylenediamine tetra-acetate (EDTA) (27 μM); composition (mM): NaCl 118, KCl 5.7, $CaCl_2$ 0.8, KH_2PO_4 2.0, $MgSO_4$ 2.0, $NaHCO_3$ 25, glucose 10 and gassed with 95% O_2–5% CO_2; Krebs' bicarbonate containing ascorbic acid (170 μM) and EDTA (27 μM) composition (mM); NaCl 134, KCl 5.0, $CaCl_2$ 2.0, KH_2PO_4 1.2, $MgSO_4$ 1.0, $NaHCO_3$ 16, glucose 10.2 gassed with 95% O_2–5% CO_2, pH 7.4), artificial CSF or buffers which are compatible with experimental design. The rapid distribution of cold liquid through the capillary bed cools the brain 'inside out' as opposed to the lengthy and non-uniform method of cooling from the surface. This simultaneous fast cooling of all brain regions makes the brain easier to section, minimizes post-mortem catalytic changes and produces a more 'transparent' slice with enhanced contrast of nuclear areas.

The perfusion of the vascular bed is performed as follows: rats usually not less than 200 g are anaesthetized with Nembutal or Equithesin* (or other anaesthetic) and transferred to a cold room. The animal is placed on a tray and the chest is cut open with a pair of scissors while holding the sternum, the thickest flap being kept out of the field of work with the help of a forceps. The ascending vena cava and descending aorta artery are clamped at their entrance to the thorax with a mosquito forceps, which restricts the perfusion to the head, thorax and upper limbs. The right atrium is cut with a small scissors so as to open the vascular circuit. Immediately after cutting the atrium a short needle (23 gauge) is introduced into the left ventricle. Avoid the use of long needles since the tip can easily reach and break the aorta. We

* Equithesin is made up in the following way: for 100 ml: (1) 0.972 g sodium pentobarbital in 11.5 ml 95% alcohol; (2) 4.25 g of chloral hydrate in 42.8 ml propylene glygol (propane, 1,2-diol); (3) 2.126 g of magnesium sulphate in 45.7 ml distilled water. Mix the above three solutions in the order 1, 2 and 3).

frequently employ needles with small 'arrow shaped' stops on the shaft, made with plastic resins or metal, to securely fit the needle in the heart during perfusion. The needle is connected to a cannula and 50 ml syringe containing the ice-cold perfusion solution. Check that no bubbles are present in the system as air droplets will obstruct small vessels preventing a satisfactory perfusion. Slowly perfuse a volume of not less than 100 ml for a 250 g rat, excessive pressure will produce rupture of small vessels in the nose.

For removal of the brain follow instructions basically as indicated in Chapter 1.

Following these guidelines it is possible to obtain an undamaged brain and brain stem. If the olfactory bulb is included in the study, the bone should be cut beyond the tip of the bulb and, with the help of a small spatula and fine scissors, the numerous branches of the olfactory nerve cut at the base of the bulb. Before removing the brain, the optic and the trigeminal nerves and also the meninges should be cut as distantly as possible from the brain with the tip of a scalpel or fine scissors. The caudal limits of the brain stem, especially the tentorium of the cerebellum, should also be carefully cut. Lift the brain gently with a spatula while cutting any adherent cranial nerve or portions of the meninges.

Once the brain has been removed from the skull place it on the plate of the tissue chopper or chuck of the Vibratome.

2.2 **Cutting sections from fresh brains**

To obtain slices of fresh brains, two instruments are recommended: the McIlwain tissue chopper and the Oxford Vibratome. The material presented in this Chapter has been exclusively done using the McIlwain tissue chopper and we include the Vibratome for the sake of completeness.

2.2.1 *The McIlwain Tissue Chopper*

Material needed (see Figure 1).

McIlwain tissue chopper (Mickle Laboratory Engineering Co) plus accessories. The original screw which holds the blade in the cutting area protrudes excessively and compresses the brain. It is therefore convenient to modify the arm and screw to prevent this.

Laboratory masking tape.

Stainless steel razor blades.

Numbered glass slides.

Fast acting adhesive. Cyanoacrylate (Loctite, IS 415, Cat. No. 41505. Ireland, or Loctite Quickset 404 Boehringer Corp. U.K.).

Beaker with ice-cold Krebs' solution (or saline).

10 ml syringe and 25–30 gauge, $1\frac{1}{2}$ in needle.

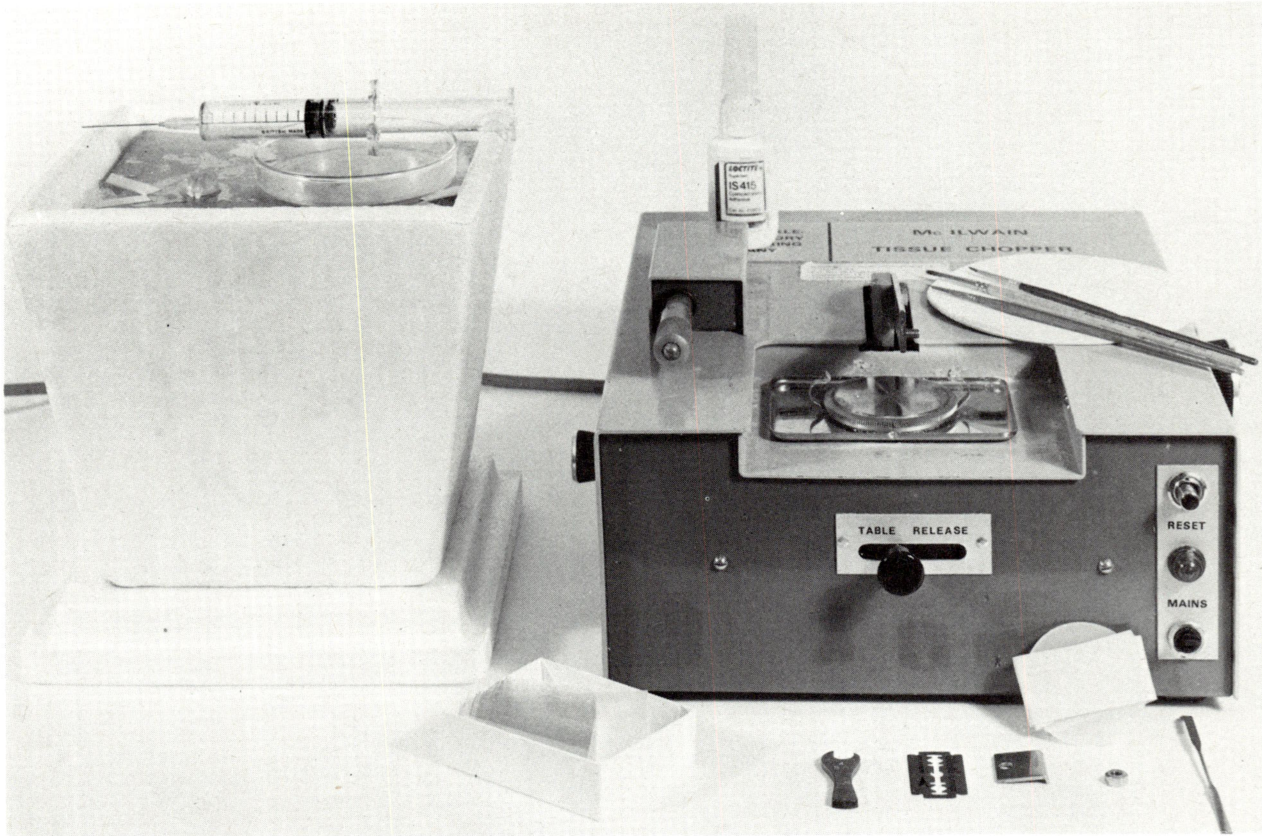

Figure 1 Materials required for obtaining fresh brain sections with a McIlwain tissue chopper

Fine brushes.
Spatula.
Absorbent paper (wipes or filter paper).
Petri dish.
Ice buckets (styrofoam/telgopor) containing crushed ice.
Metal plate which can be placed on top of ice in ice bucket.

Procedure

Before perfusing the animal, the tissue chopper should be prepared as follows.

1. Add several layers of laboratory masking tape to the original plate, this facilitates completeness of section through brain.
2. Adjust the new razor blade to the arm of the instrument checking that the blade cuts the first layer of tape evenly along most of the width of the tape. The blade has to be adjusted firmly on the arm, but not excessively so, as exaggerated pressure on the blade bends it and produces irregularly thick sections.
3. Preset advance knob for desired thickness.
4. Set lateral knobs for maximum force and minimum velocity.
5. Fill syringe and petri dish with cold Krebs' solution.

Cutting tissue slices

1. Gently blot the surface of the brain to be glued to the plate. For most areas the brain is glued to the plate with ventral aspect upwards. However, as the glued surface is easily torn during sectioning it is advisable to place the more relevant areas upwards, facing the blade. When the brain stem is the object of microdissection it is convenient to previously remove the cerebellum by cutting its peduncles. When only one operator performs all the manipulations including the microdissection of specific nuclear areas, usually not more than ten to 15 sections per rat brain can be conveniently handled. Larger samples require at least two operators: one producing and transporting the tissue sections and the other microdissecting the slices under the microscope.
2. Spread a thin layer of Loctite (or equivalent adhesive) over the tape, covering approximately the area that will be occupied by the brain. Excessive adhesive will spoil the blade.
3. Place the brain on the area of the tape spread with adhesive. Orient the brain as perpendicular as possible to the cutting arm to ensure maximum symmetry in the sections (Figure 2A). Ensure proper contact between the brain, tape and adhesive by placing a wet glass slide (or any similar lightweight object) over the brain. Let the brain stand for three to five minutes.

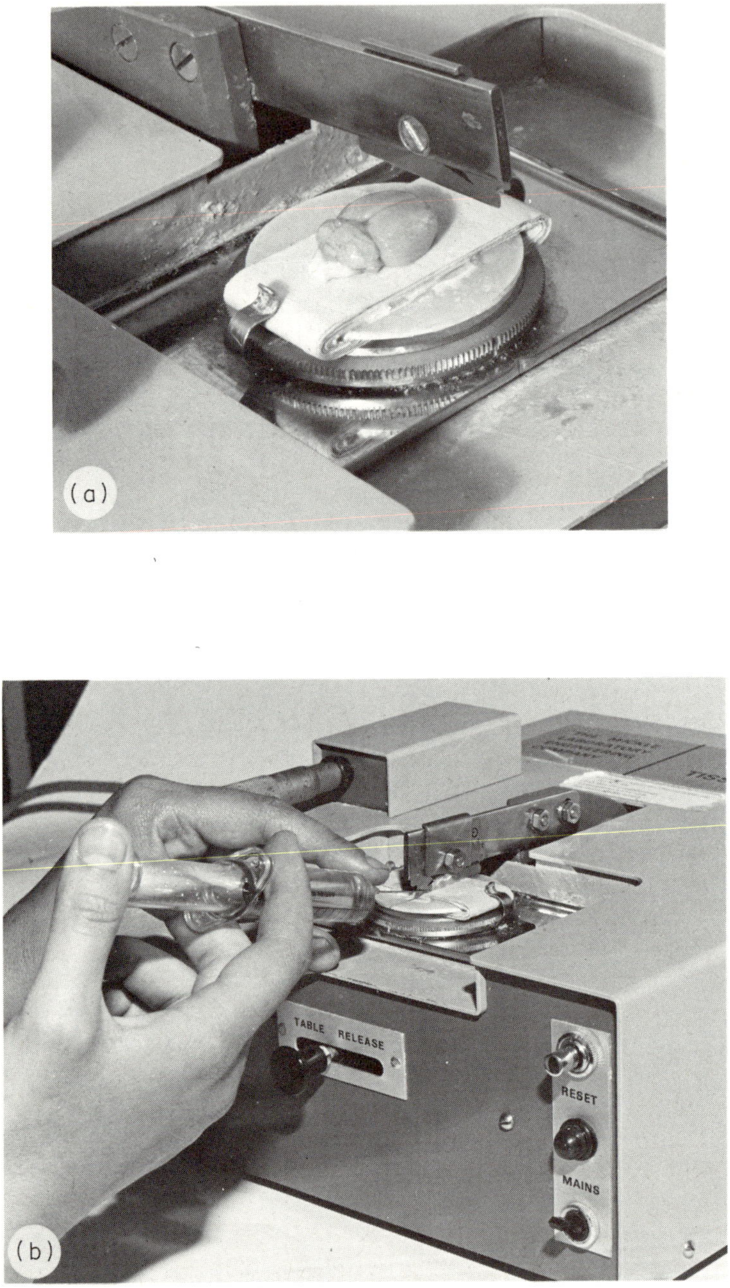
(a)
TABLE RELEASE
RESET
MAINS
(b)

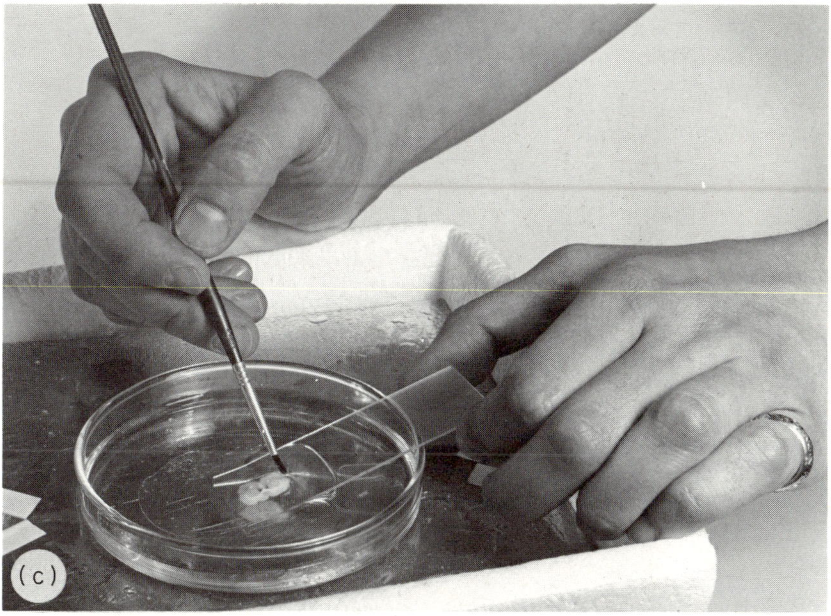

Figure 2

4. To start cutting, advance the plate manually to position the brain near the blade. If the relevant areas are not present in the initial sections set the advance to 500 μm or more and remove the unwanted sections with a small spatula. When approaching the areas of interest reset the advance knob to 200–300 μm (or less if desired) and start collecting two or three sections prior to the area of study. While sectioning keep the blade clean of adhesive and moist with Krebs'. Avoid excessive liquid on the side of the tape where the brain is placed.
5. The sections are collected in the following manner. Advance the blade one step manually until it falls with maximum strength. The blade should cut the brain evenly and completely. While the brain is still with the cutting edge on the tape (see Figure 2B) and effectively dividing it into two compartments, insert the tip of the needle between the blade and the section with the bevel pointing towards the blade. At the end of this operation the section should hang over the needle. This step may require some practice.
6. Float the section in a petri dish containing ice-cold Krebs' solution. (Dry the tape before lifting the blade to obtain the next section.) Collect the

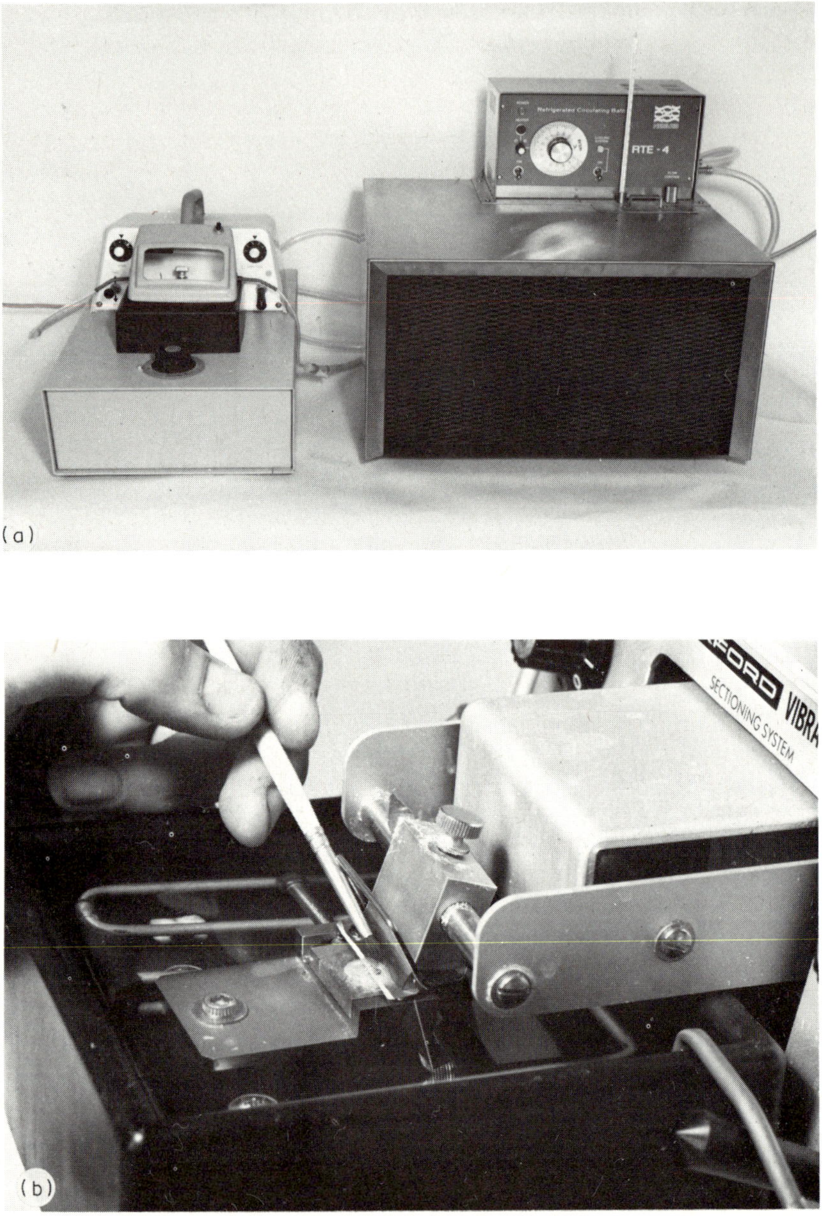

Figure 3 (a) The Oxford Vibratome equipped with a stainless steel serpentine linked to a thermostatically controlled cooling system.
(b) The vibrating blade of the Vibratome has completed a 100 μm thick section of the fresh rat brain. The freely floating slice is collected with a fine brush. See the disposition of cooling serpentine around the tissue holder

section on a numbered glass slide with the help of a fine brush as indicated in Figure 2C. Ideally only one section should be placed on a single slide. More sections can be collected if only one particular nucleus is being microdissected in several consecutive sections. If the side (left-right) is relevant it is advisable to make a cortical indentation (or any other convenient mark to allow side identification) prior to placing the brain in the tissue chopper.

7. The sections are placed in the corresponding order over the metal plate in the ice bucket. The distance of this plate to the top should be such as to permit covering it with the lid to transport the sections to the laboratory.

2.2.2 *The Oxford Vibratome*

Materials

Vibratome (Oxford Instruments, Cat. No. 501 and 502) and accessories.
Blades (original Vibratome or conventional razor blades cut in half).
Adhesive: Cyanoacrylate (Loctite, IS 415, Cat. No. 41505. Ireland, or Loctite Quickset 404 Boehringer Corp. U.K.).
Fine brushes.
Fine forceps.
Petri dish.
700 ml Krebs solution (at least 600 ml required to fill bath).

Procedure

To keep the Vibratome bath cool it is convenient to place a serpentine of stainless steel tubing connected to a pump assisted cooler (Neslab RTE-4) in the bath. The Vibratome–cooler bath set up is illustrated in Figure 3A.

The Vibratome allows the preparation of thinner slices of fresh, unfixed tissue with good structural preservation. The limitation of this approach is that only portions of the rat brain can be cut on each operation and that the procedure is more time consuming than with the tissue chopper. Nevertheless, while we recommend the use of the tissue chopper for most neurochemical studies involving a large number of samples, the use of the Vibratome might prove invaluable for a number of specific applications, i.e. when preparing sections of fresh tissue for electrophysiology or for neurotransmitter uptake-release of a single nuclear area. We have used such slices to study uptake-release of 3H-dopamine by electrical stimulation of the substantia nigra and caudate putamen (Aceves and Cuello, 1981).

Cutting tissue slices

1. Fill the bath of the Vibratome with cold Krebs' or other buffer solution or cool the bath prior to sectioning. Good sections can be obtained up to 10 °C but for best results aim to keep the bath temperature between 2 °C

to 4 °C. If a serpentine is used, the circulating coolant should be set below zero to obtain a temperature slightly above zero in the bath. The operator should find the optimum low temperature setting which, while keeping the bath cold, does not freeze the solution around the tubing thus preventing temperature exchange between the bath and coolant.
2. Place a new blade for each tissue block and set angle between 15 ° to 25 °.
3. The brain must be freed of any meningeal cover and has to be trimmed to an adequate shape to accommodate the Vibratome chuck. Brain stem and most brain areas can easily be prepared for sectioning. The olfactory bulbs present a special difficulty as the two bulbs are not 'supported' by other brain structures. Nevertheless, successful sections of this structure can be obtained by either cutting the bulbs longitudinally and firmly adhering them to the chuck or by placing the bulbs with a portion of the frontal pole cut at an angle on the chuck with the bulbs facing the blade.
4. As for sections with the tissue chopper, spread a minimal amount of adhesive over the clean surface of the chuck. Reset the chuck holder to the lowest position to allow maximum range of advance. Orient the tissue and chuck in relation to the blade. Bring tissue 'upwards' to a position almost touching the blade. Set equipment for maximum vibration and low velocity, usually position 2–4.
5. To obtain an even surface, cut two or three thick slices in the range 300–400 μm. Discard these slices. Once an even surface has been obtained reset advance for thinner slices, 250–100 μm (or less) depending upon need. Slices of entire transverse sections of fresh rat brain cannot be obtained easily below 100 μm. Sometimes it is necessary to support the tissue with a brush at the final edge of the section pushing it very gently towards the vibrating blade to complete the section. This has to be done with care in order not to displace the rest of the tissue block. A 100 μm thick Vibratome section of the entire rat brain is illustrated in Figure 3B. In order to obtain sections of consistent thickness it is recommended that the blade holder be moved backwards just beyond the tissue before beginning the process again. With the aid of a brush, collect the tissue slice on a glass slide and place it on a metal plate over crushed ice as described in Section 2.2.1.

3. MICRODISSECTION OF FRESH BRAIN SLICES

3.1 Materials

1. Dissecting microscope with a transillumination system (optimally with accessories for microphotography).
2. Cold stage.
3. Dry ice containers.

4. Microknife (Castroviejo scalpel, iridectomy scalpel or laboratory made instruments).
5. Dissecting needles (skin graft, histology needles or laboratory made instruments).
6. Cold metal plates.
7. Anatomical atlases.

As the tissue slices have to be kept cold during microdissection a cold stage which allows the passage of light is necessary. For this Zigmond and Ben Ari

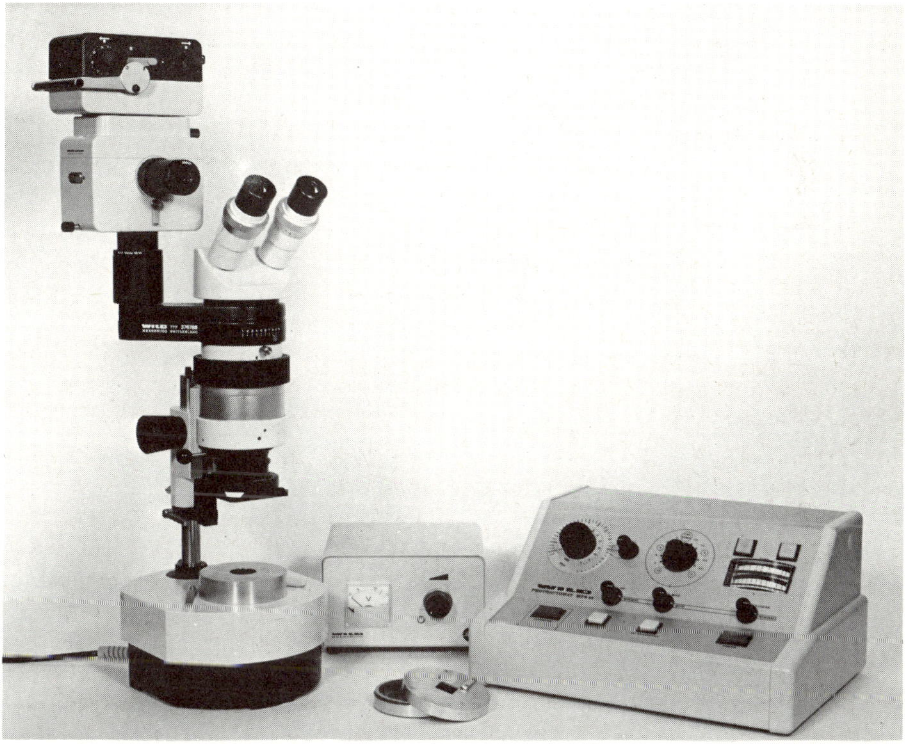

Figure 4 Dissecting microscope equipped with transillumination and automated system of micrography. The base of the microscope contains light source, diffuser and fan. Note the laboratory-made cold stage fitted on to the microscope and standing on the table

constructed a metal stage composed of two hollow matching rings to lodge dry ice in its interior (see Figure 4). These stages have to be built to fit the corresponding microscope. Although most commercially available stereo microscopes are adequate for this procedure, those with better resolution and provided with a transillumination system will considerably facilitate the

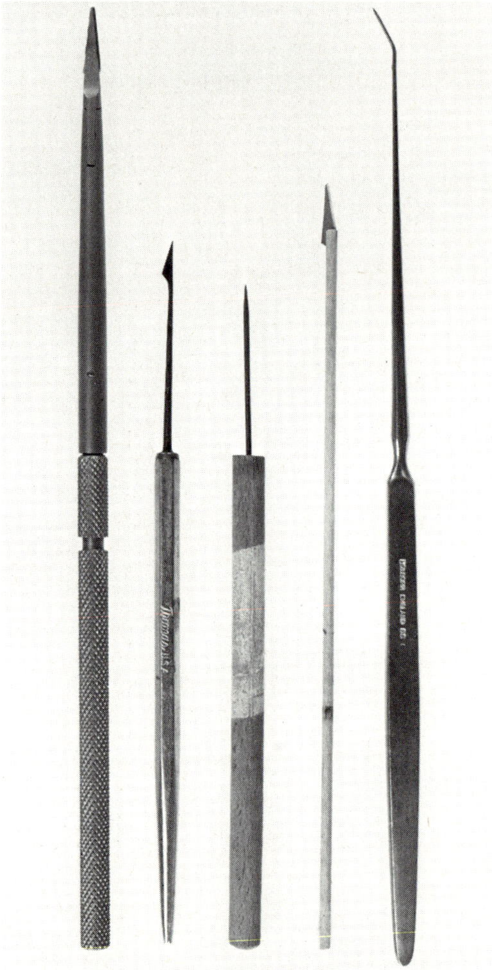

Figure 5 From left to right: Castroviejo scalpel, iridectomy scalpel, histology needle, cut piece of razor blade glued to a wooden stick, and skin graft needle

operation. An ideal instrument should have good resolution, transillumination, a wide field, a large focal distance, a wide range of magnification and a automated system of microphotography. Of these conditions transillumination is the most important as the procedure described is based on the recognition of neuroanatomical structures by their differential transparency. We have found the Wild M7S microscope particularly useful for microdissection. Transillumination in this microscope does not produce heat on the stage as the light source is relatively distant and is fan cooled. When transillumination is not present in the stereomicroscope this can be provided by a home-made transilluminated stage. These should also be provided with a diffuser as systems based only on mirrors produce very uneven illumination. The lamp should be placed at a reasonable distance from the stage to avoid heating and its source of power should be variable so that optimal illumination for a particular area in a given slice can be attained. Stereomicroscopes with a wide field allow one to work with a high magnification while still having reference to the over-all organization of the slice. A large focal distance permits easy manipulation of the dissecting instruments under the objective while viewing through the eye pieces. Finally, a convenient system of microphotography will facilitate the documentation of the microdissections as well as establishing landmarks for future sampling.

The basic tools needed for microdissection of fresh slices are a fine needle and a microknife (Figure 5). Histology needles are adequate although skin graft needles (Holborn. Cat. No. BN374) are heavier, longer and provide a better fit to the hand. A variety of microknives can be used for microdissection of fresh tissue slices. The simplest knife is made with cut pieces of razor glued to a long wooden stick with epoxy resin. This is not only economical but also has the advantage that a large selection of knives can be constructed with blades of different shapes and angles depending on their use. The Castroviejo scalpel (Moria-Dugast, Paris. Cat. No. 9927) can be modified in the same way, but has the additional advantage that the tip can be accurately guided like a pencil and its shape and weight are optimal for long microdissection sessions.

If the purpose of the microdissection is to obtain samples for biochemical analysis, the microdissected material is collected in correspondingly labelled test tubes. If the samples are to be weighed or pooled, it is more convenient to recover them in excavated glass slides, a properly numbered black adhesive tape on the opposite side of the slide permits a better visualization of the samples. If the purpose of the microdissection is to obtain samples for studies on the uptake and/or release of neurotransmitters it is essential to avoid freezing the tissue, for this the excavated glass slides are kept with a drop of Krebs' or suitable solution on a cold metal plate.

3.2 **Procedure**

Here we intend to offer some broad guidelines on how to approach the microdissection of nuclear areas and fibre tracts of the rat CNS. Investigators should adapt these recommendations to fit their needs and laboratory facilities.

Once all the necessary materials are prepared, the operator can begin the microdissection of the desired area of the CNS with the aid of an atlas. For the rat brain the atlas of König and Klippel (1963) is the most suitable as the plane of sectioning is closer to the natural plane obtained by simply placing the brain with the cortex downwards on a tissue chopper. This atlas, however, finishes at mesencephalic levels and thus for more caudal levels the atlas of Palkovits and Jacobowitz (1974) should be used. Sections obtained with a tissue chopper or a Vibratome without stereotaxic landmarks result in planes which are approximately similar but not identical to those of the above atlases. The most common difficulty which arises when cutting sections on the tissue chopper is that the dorsal aspects of the slice correspond to a different level than those of the ventral aspects as represented in the above atlases. Asymmetry is also a frequent problem when the brain has not been placed perpendicularly to the blade. It is therefore of paramount importance to become familiar with the appearance of nuclear areas and tracts by the microscopic inspection of a large number of sections before sampling for actual experimental purposes. Although each section will be slightly different from the equivalent plane in the atlas, it is possible with some practice to follow the features of most nuclei and fibre tracts. As an aid in the recognition of some classical nuclei and fibre tracts a collection of micrographs from representative slices is reproduced in Section 4.

There are two ways of identifying the different structures in fresh brain slices. One is the method of transillumination employed in our laboratory which permits identification simply based on the differential optical density of myelinated tracts (dark) and neuronal cell bodies (clear). The other is that of Zigmond and Ben-Ari (1976) who stained the tissue in 0.2% methylene blue in order to identify the different brain areas. Thus cell rich areas would appear heavily stained while fibre rich areas would appear lightly stained. Briefly, their method is as follows: a 0.2% solution of methylene blue is made up in the appropriate buffer and left stirring overnight before use since methylene blue often precipitates out upon standing. Sections are stained for about ten seconds at room temperature and then washed twice in ice-cold buffer. This light staining affects only the most superficial layers and does not affect a number of biochemical parameters such as choline acetyltransferase, glutamate decarboxylase or tyrosine hydroxylase activities, nor does it interfere with protein determination by the Lowry method. It would, of course, be necessary to perform control experiments to determine that methylene blue staining did not interfere with other parameters under study.

During microdissection the slices should be kept moist (but not excessively so) and as cold as possible. Only a limited number should be brought from the cold room to the laboratory where they should be kept on a metal plate in an ice filled bucket. The bucket should then be covered (see Section 2 for explanation). Only one slide at a time is removed from the metal plate and brought to the cold stage. The stage is partially filled with dry ice in order to keep the slide cold but not frozen. If crystals start to form on the slide, remove it immediately from the stage and return it to the metal plate.

The main steps in the microdissection of a particular brain area are as follows:

1. Identify all the relevant structures of the slice at low magnification (5× to 10×). Regulate illumination to obtain maximum contrast for the particular slice. This varies from section to section according to the amount of myelinated structure.
2. Set the microscope to higher magnification if needed.
3. Cut around the perimeter of a given structure by performing a series of firm, long strokes while holding the slice in place with the needle. The tip can be placed on a heavily myelinated structure to hold the section. Alternatively, the needle can be placed in any part of the perimeter of the area under dissection thereby supporting the microknife with the needle while cutting. This will allow the knife to perform clean cuts without disturbing the rest of the section.
4. When numerous samples are required from the same slice it is always advisable to start the dissection from the periphery to the centre. This is in order to have a central core of solid tissue which can be held. The opposite approach produces 'hollow' slices which easily lose the anatomical relations between nuclei and also become difficult to handle.
5. The microdissected samples are lifted out by placing the needle underneath the cut out piece of tissue while preventing its displacement against the rest of the slice, or if this is not possible, by being supported against the microknife.

The main difficulty of these operations is to get accustomed to the use of needle and knife under the microscope where only the tip of these two instruments are visualized. However, with sufficient practice a good visual–manual co-ordination can be obtained.

4. MICROGRAPHS OF FRESH RAT BRAIN SLICES

4.1 Description of micrographs

In this section we present a series of micrographs (Figure, 6–35) of representative areas of the rat brain prepared and photographed as described in Section 2.2. Although these slices do not correspond to planes of any stereotaxic atlas, they show that good identification of brain areas can be achieved on fresh, unfixed tissue based on the differential optical density of myelinated and unmyelinated areas. The most clearly identifiable areas at low magnifications are presented in the accompanying plates. The microscopic observation shows more detail than that revealed with conventional microphotography on 35 mm film (Ilford Pan F 18 DIN/50 ASA and Ilford

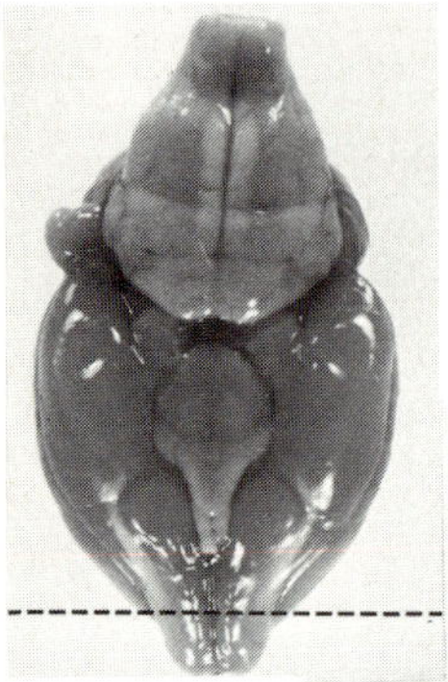

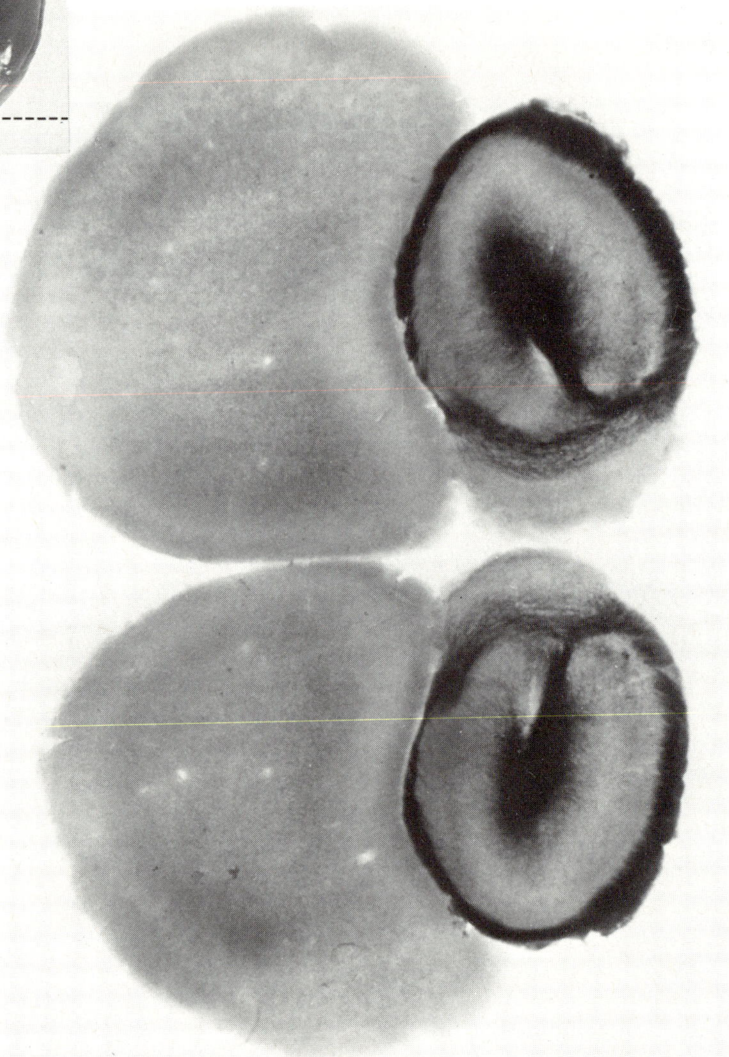

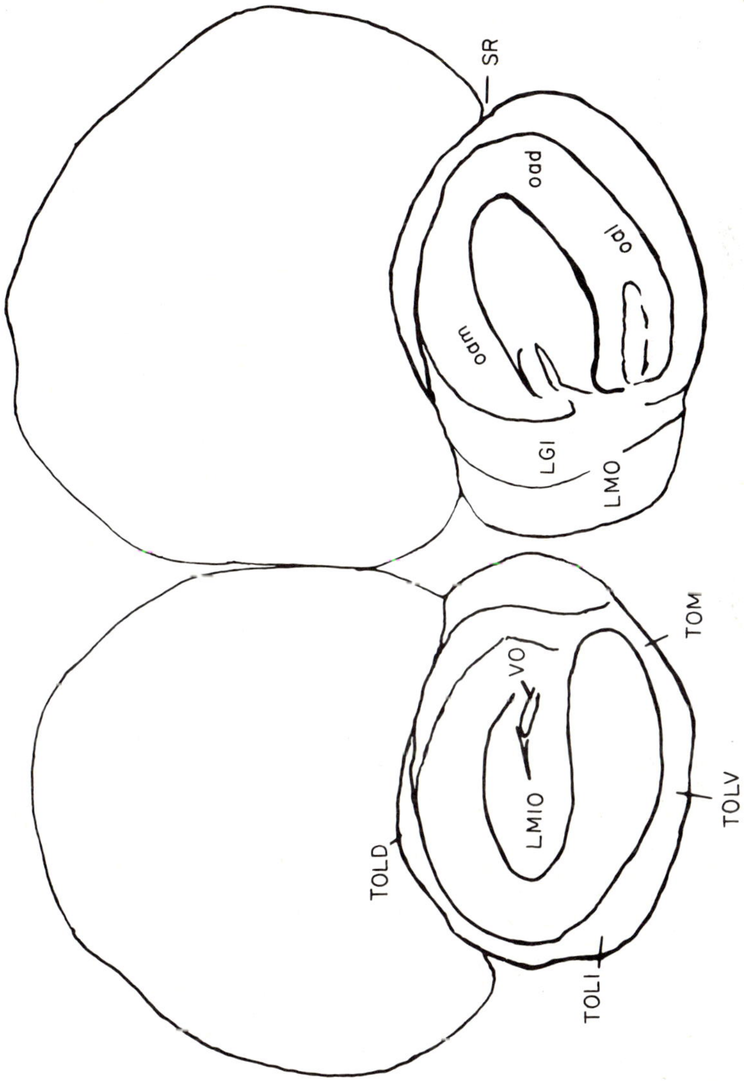

Figure 6

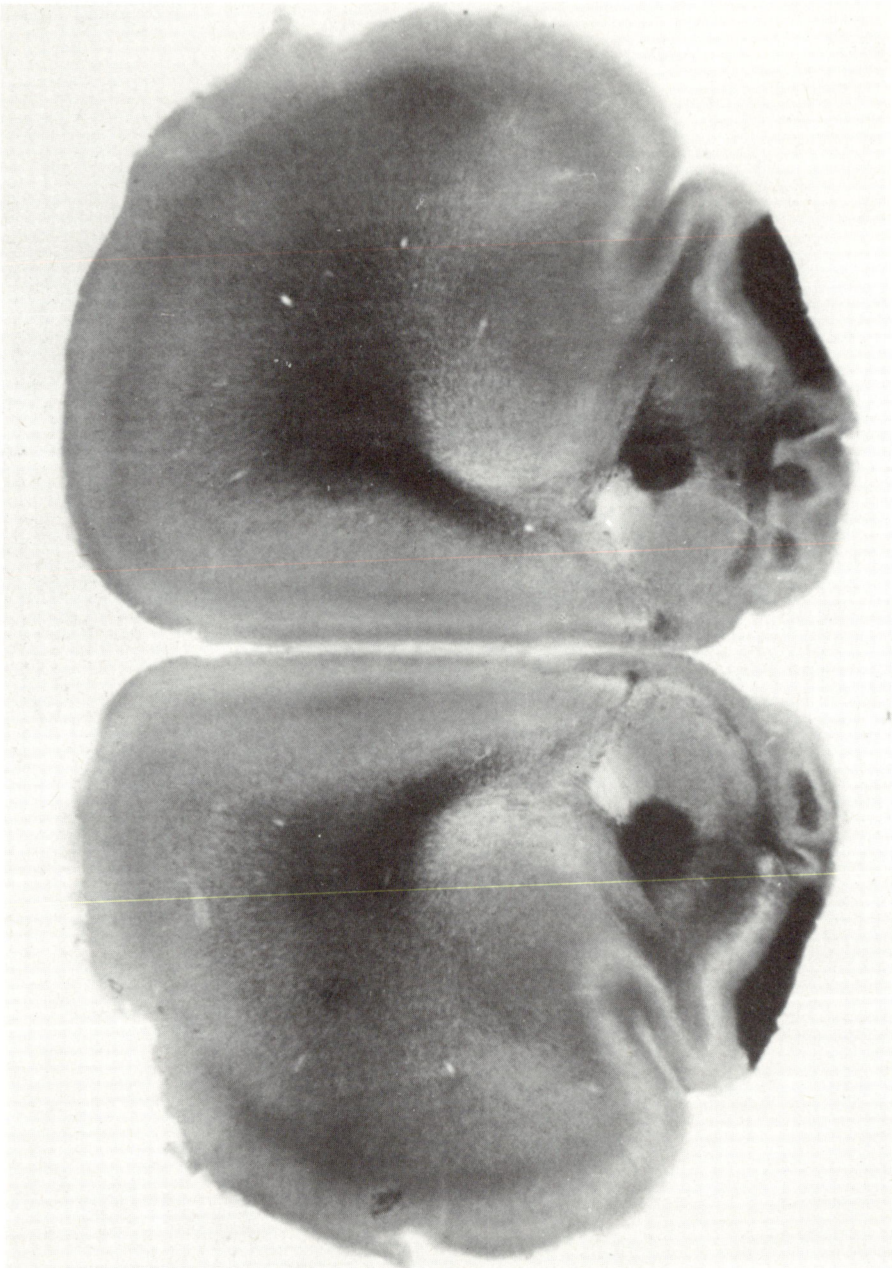

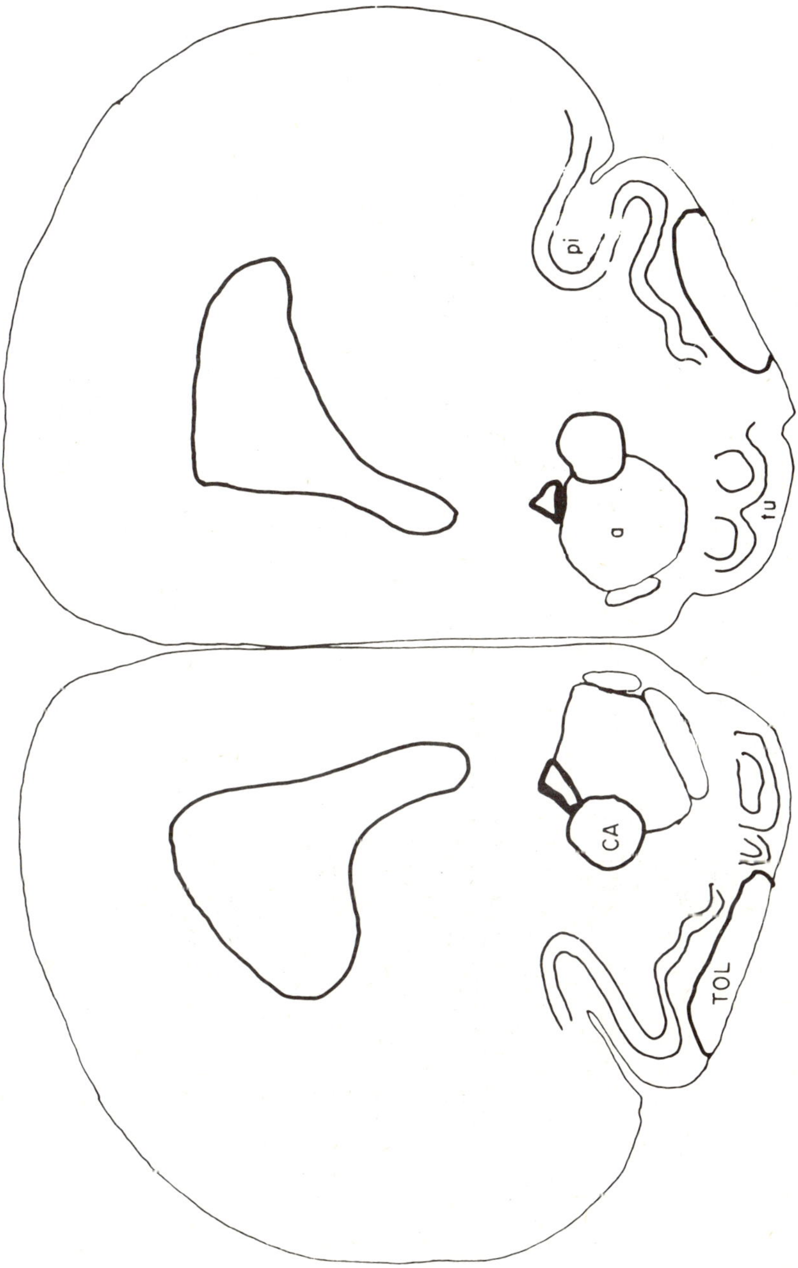

Figure 7

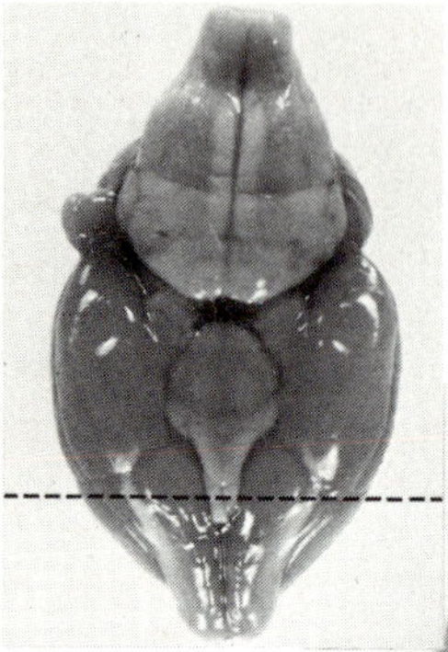

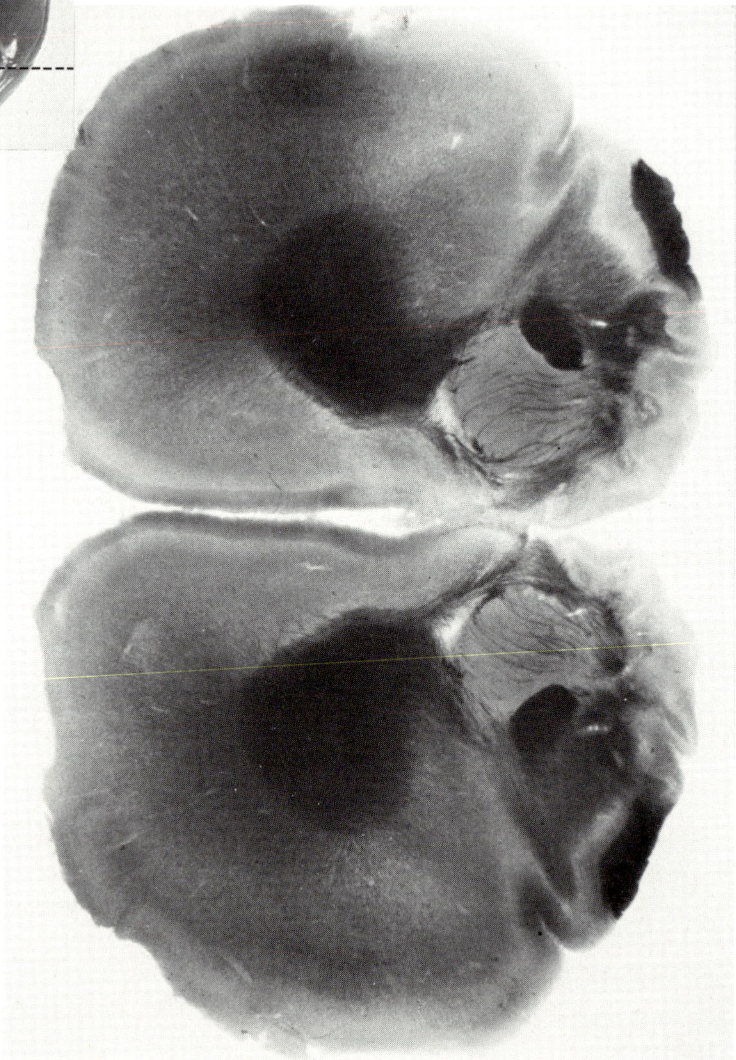

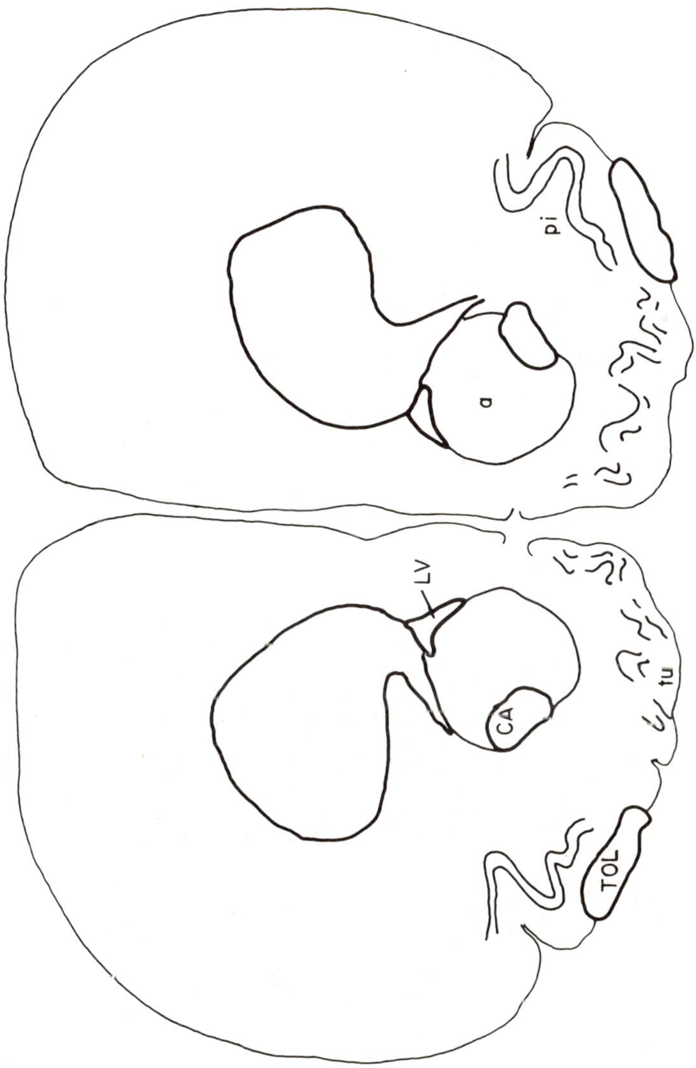

Figure 8

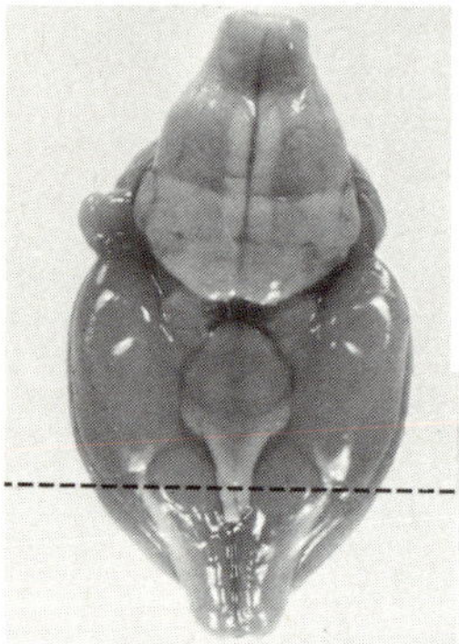

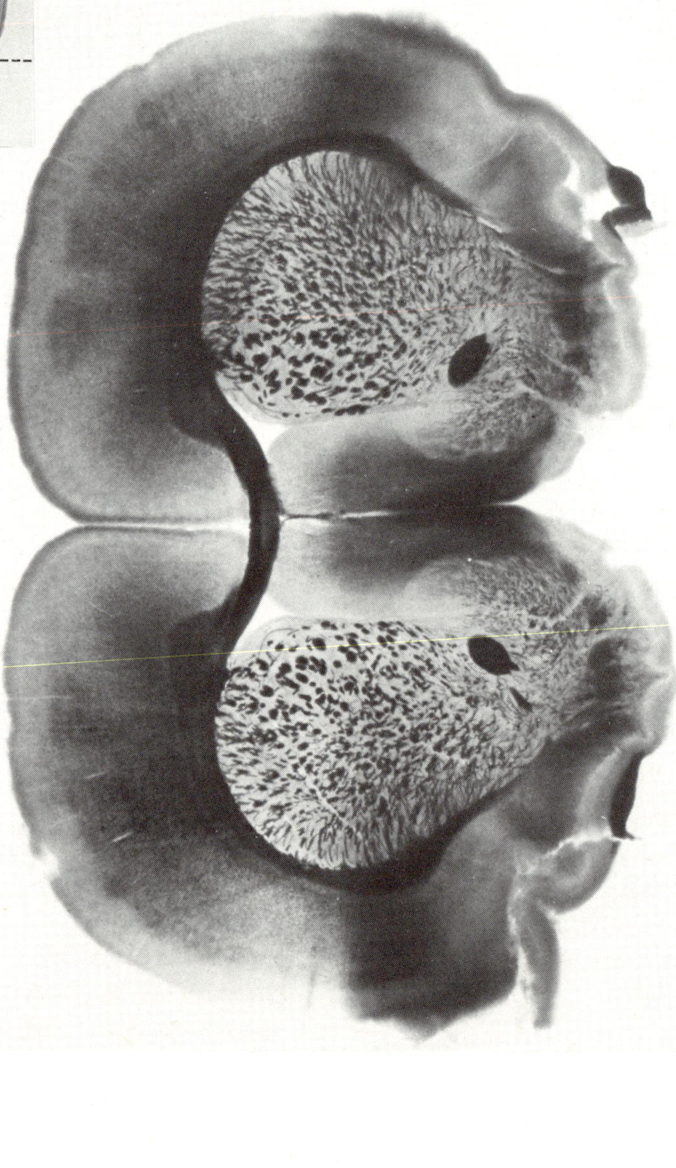

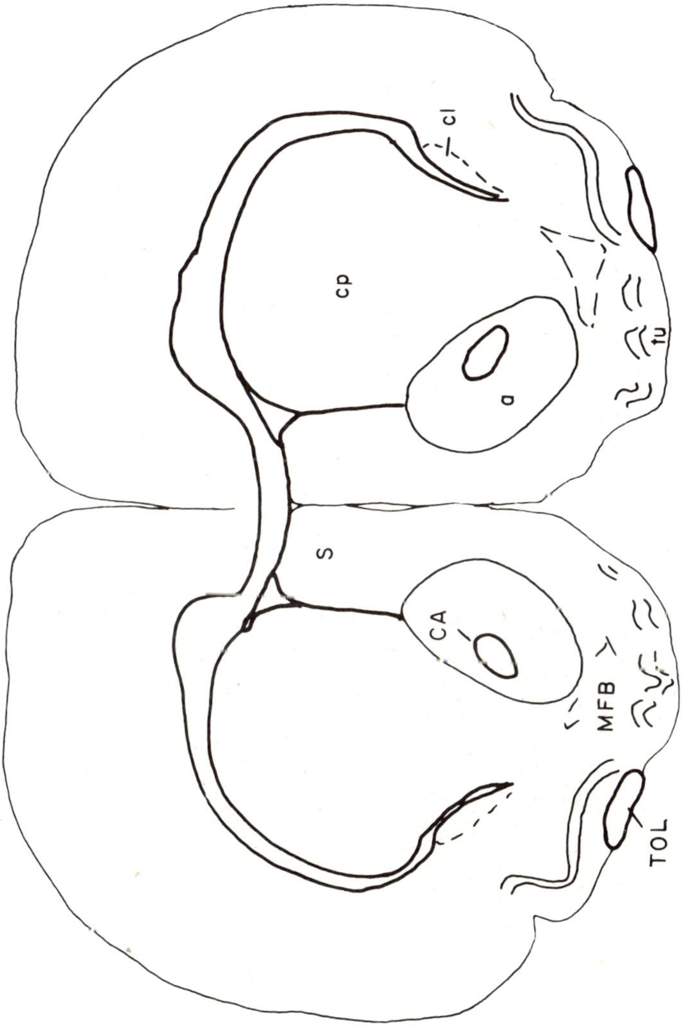

Figure 9

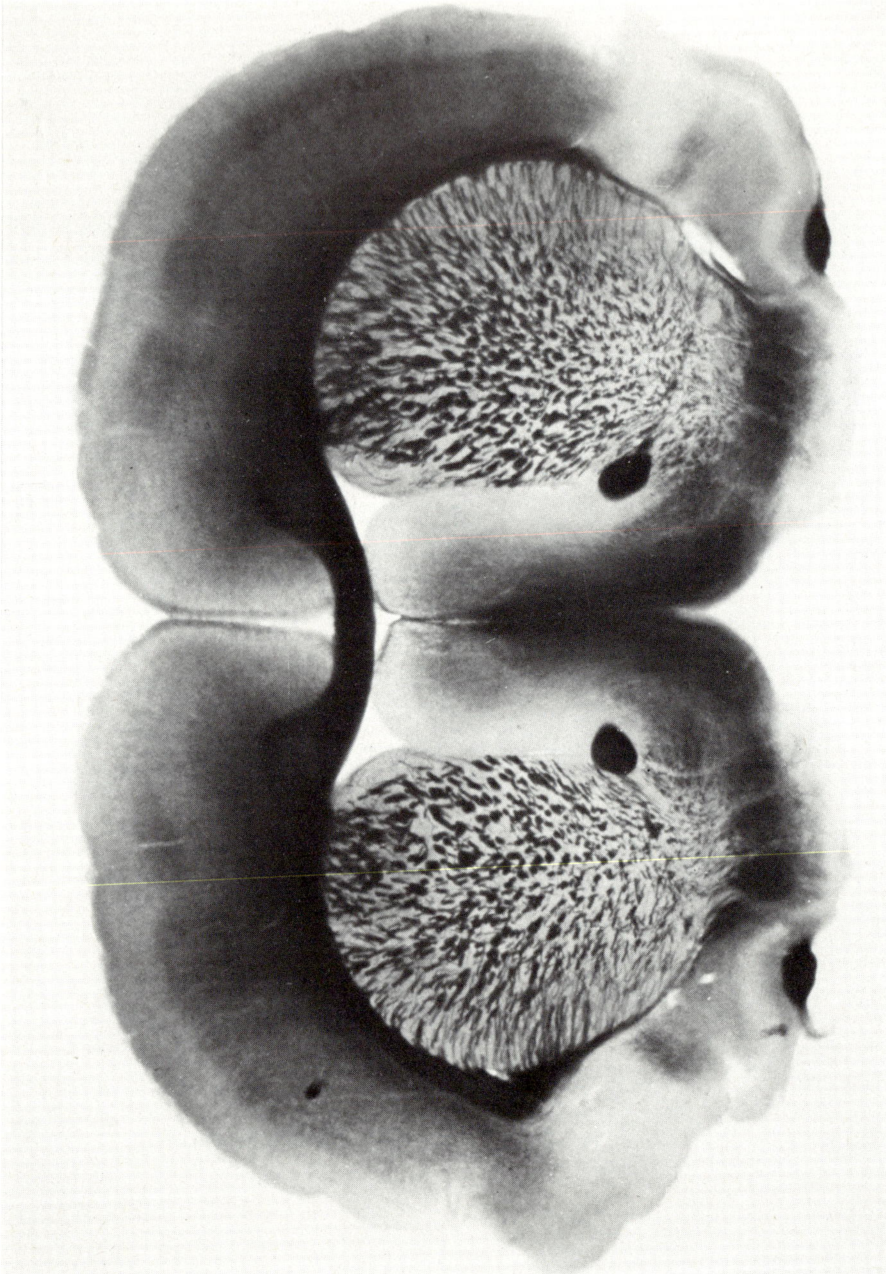

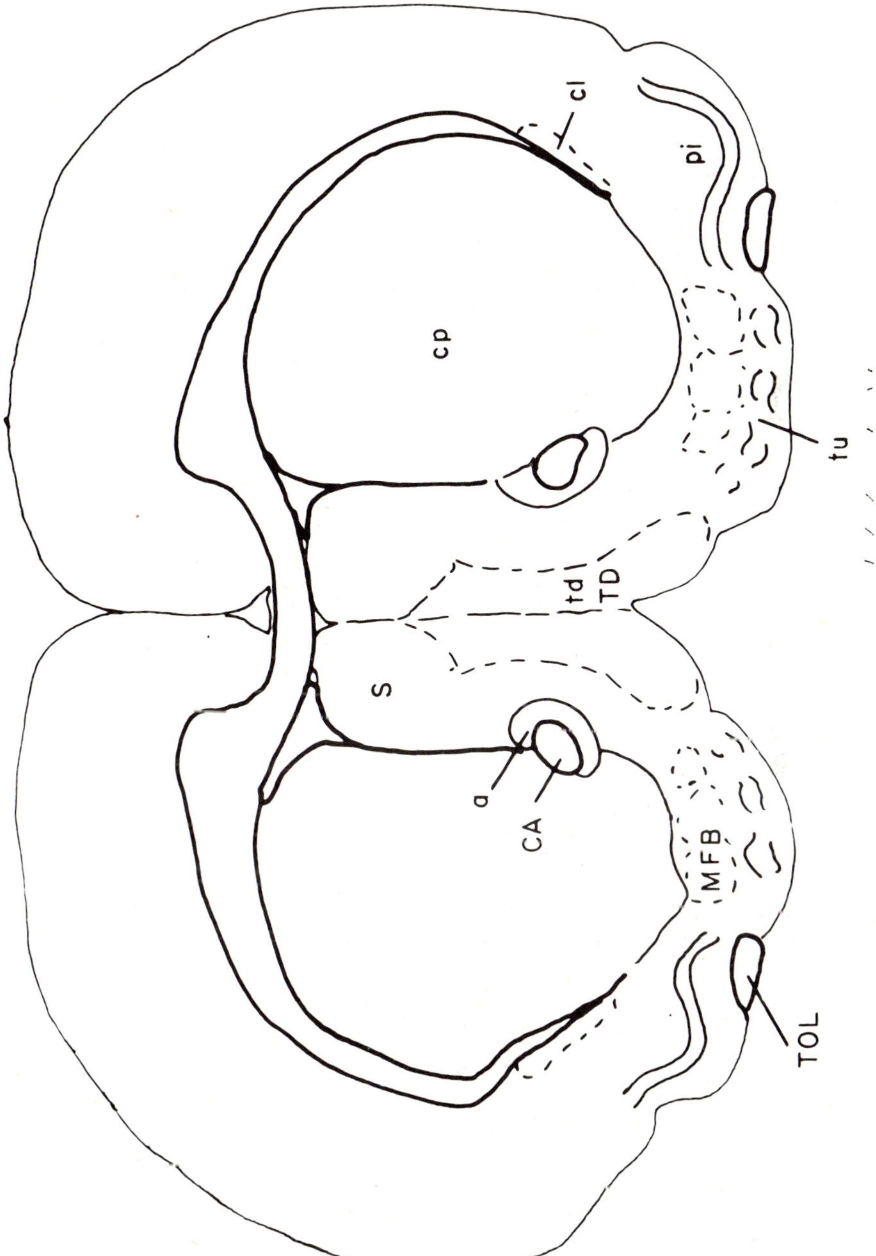

Figure 10

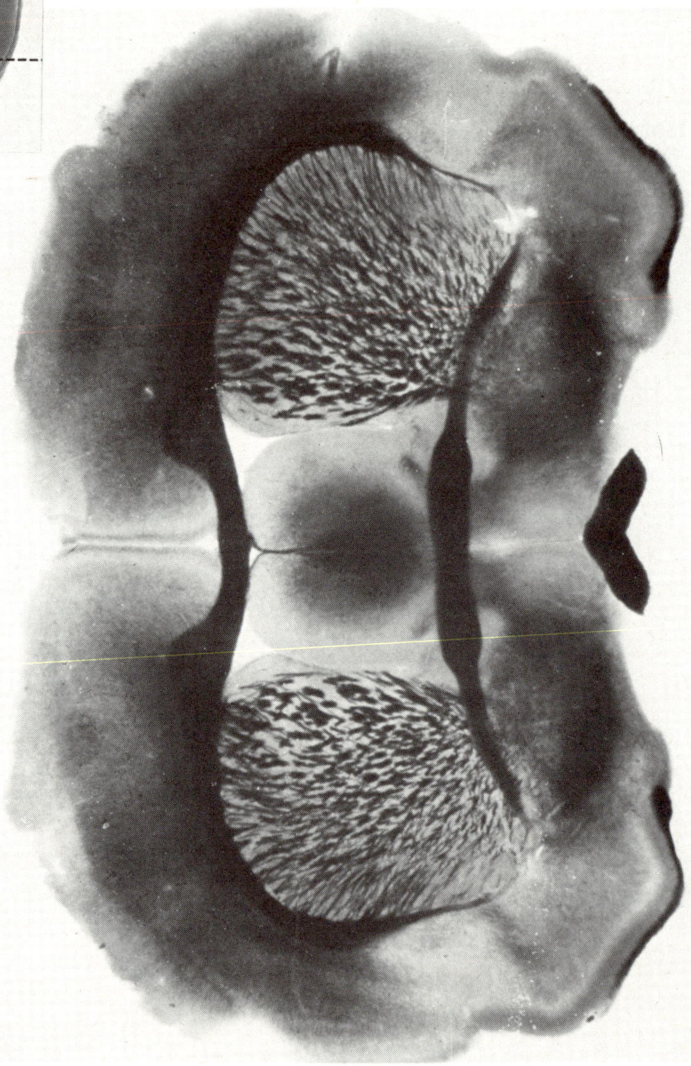

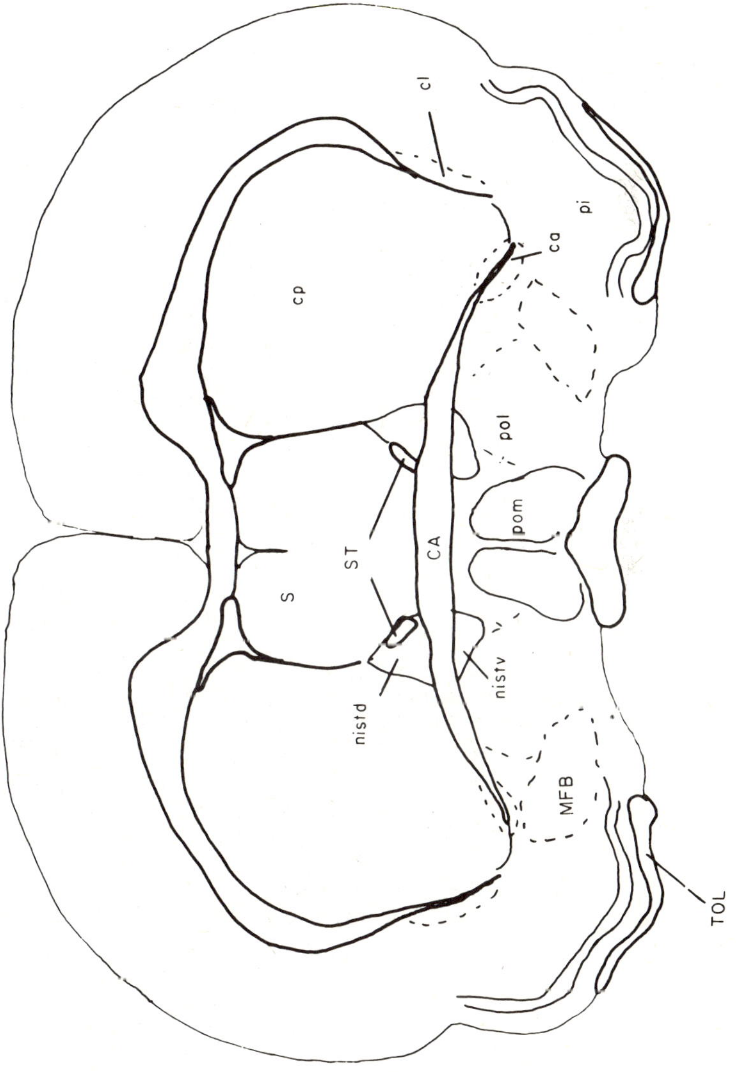

Figure 11

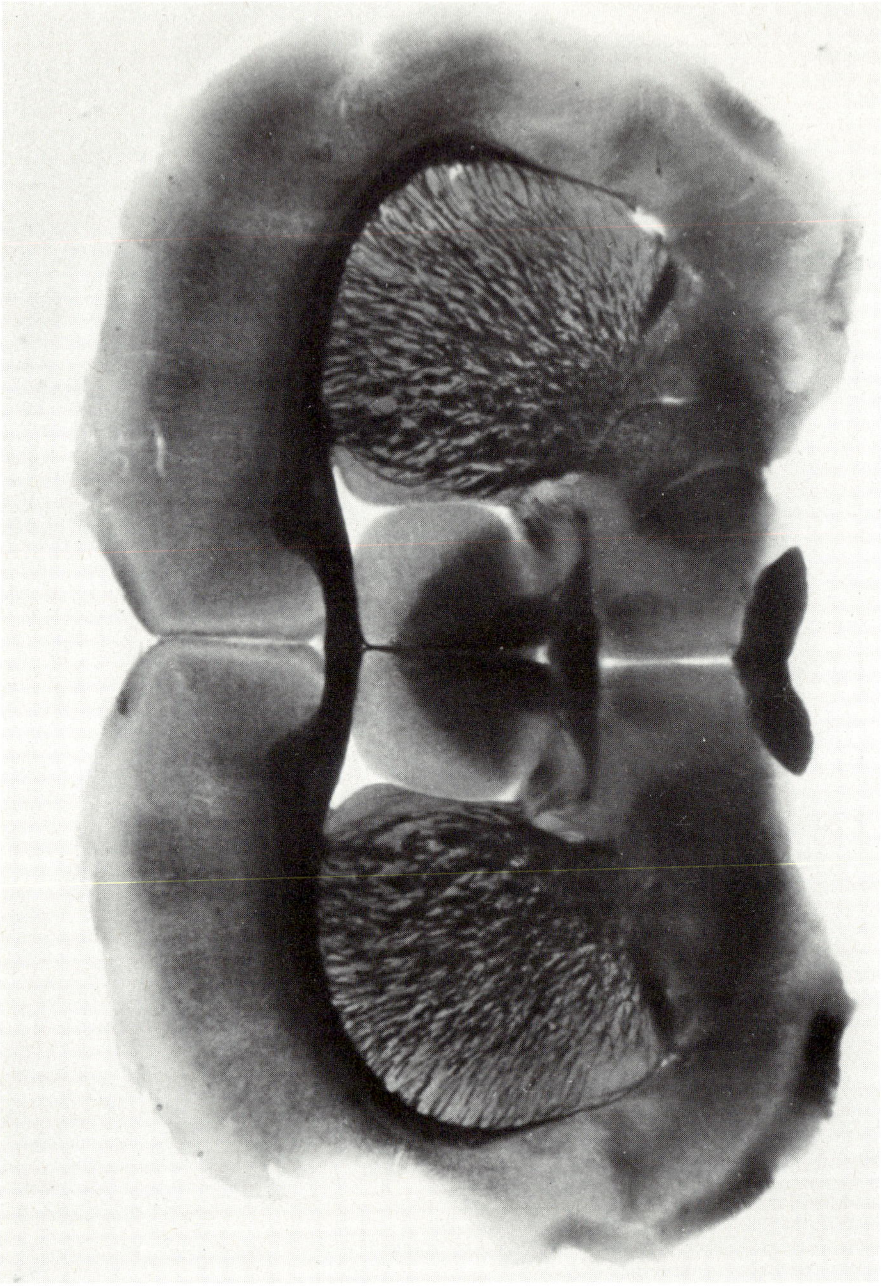

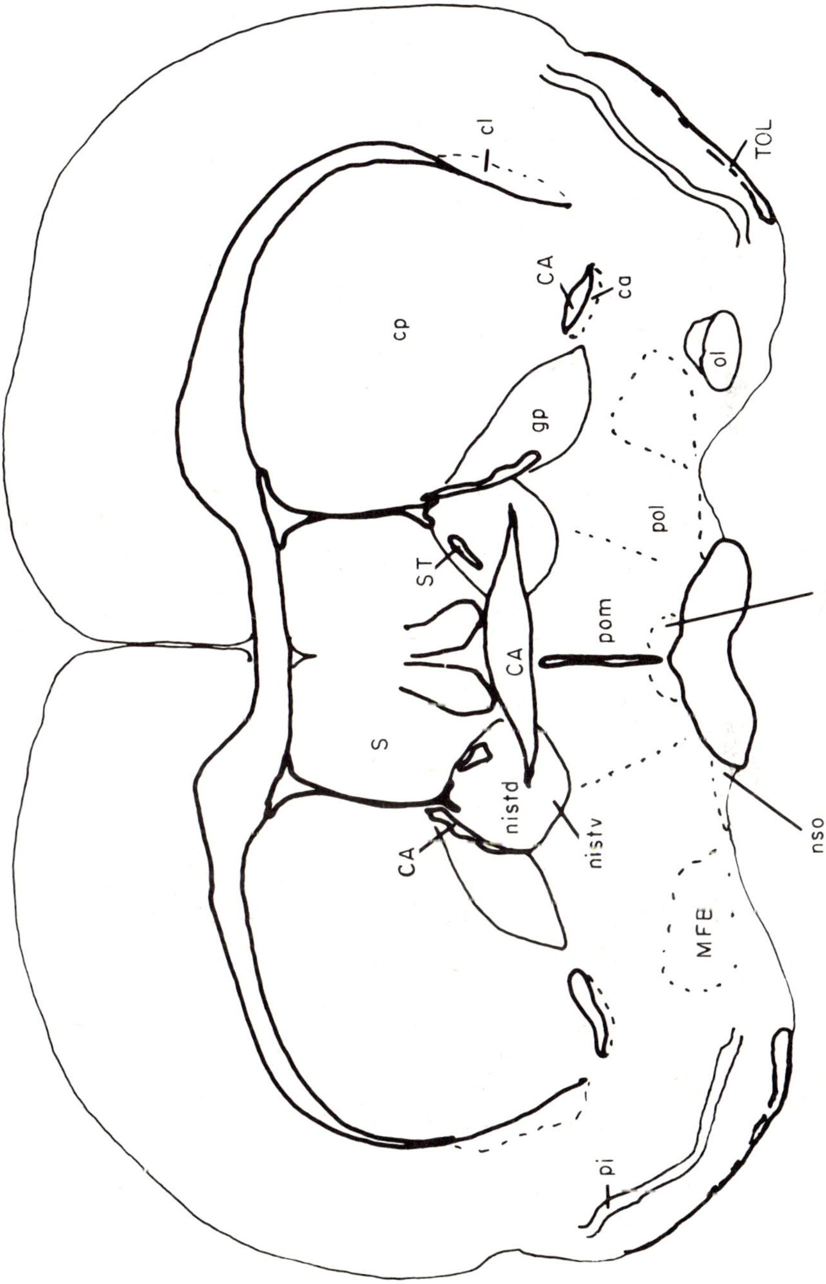

Figure 12

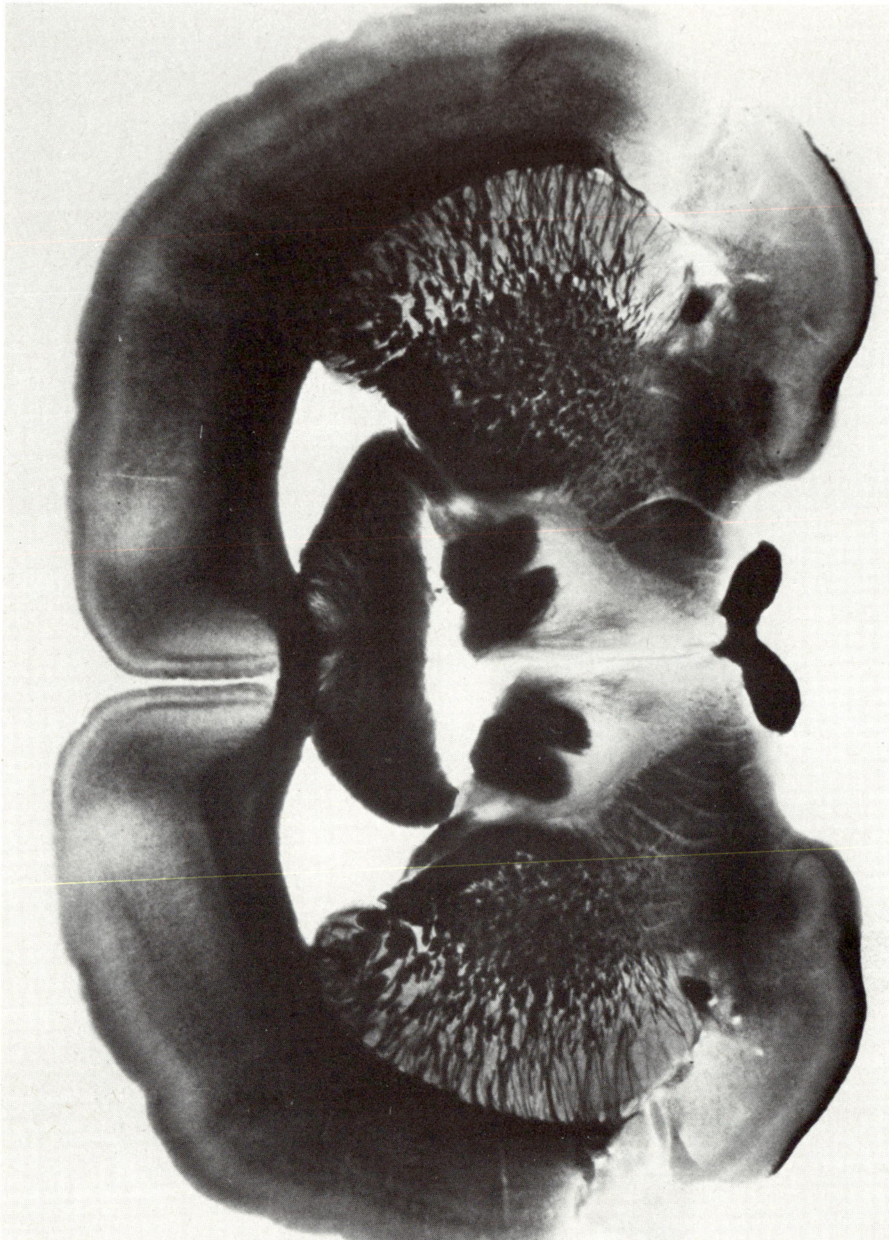

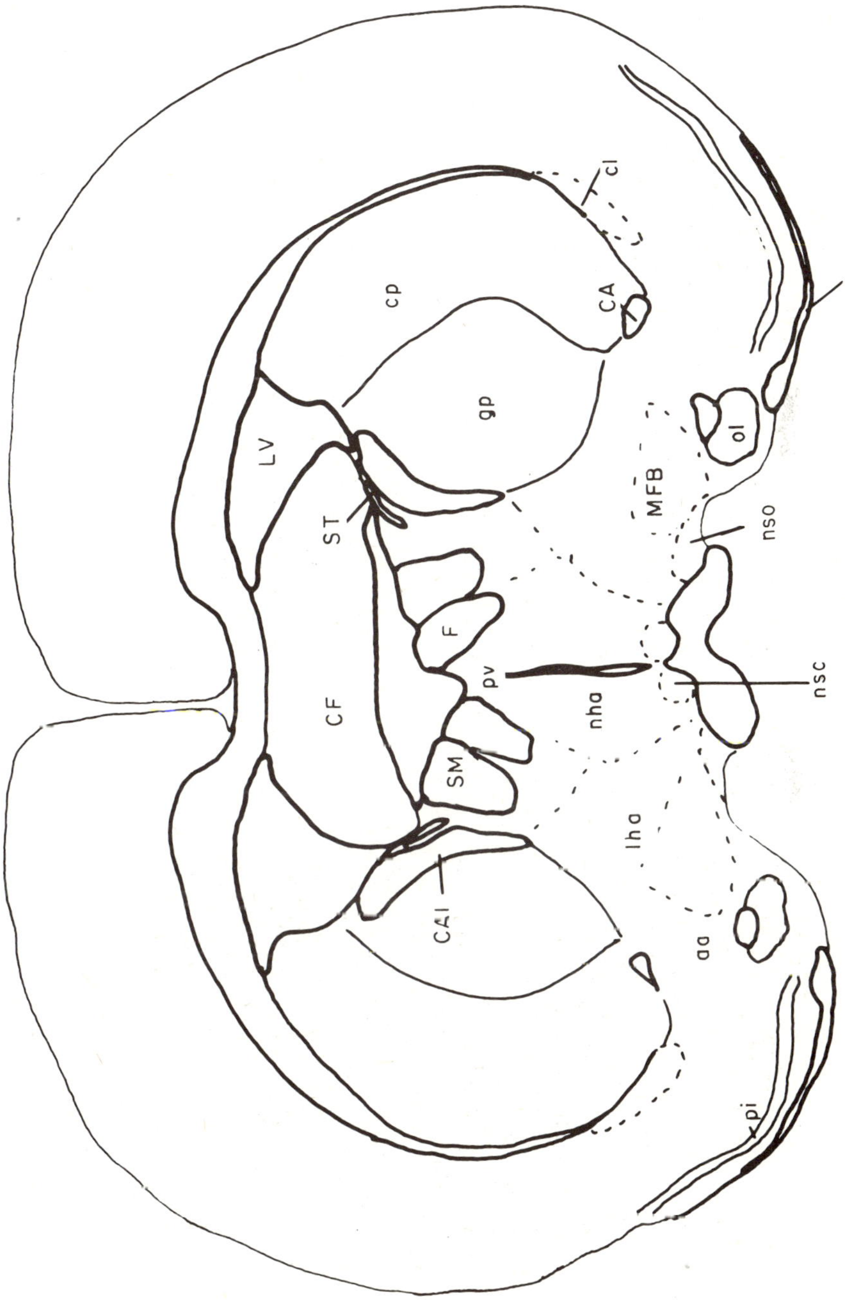

Figure 13

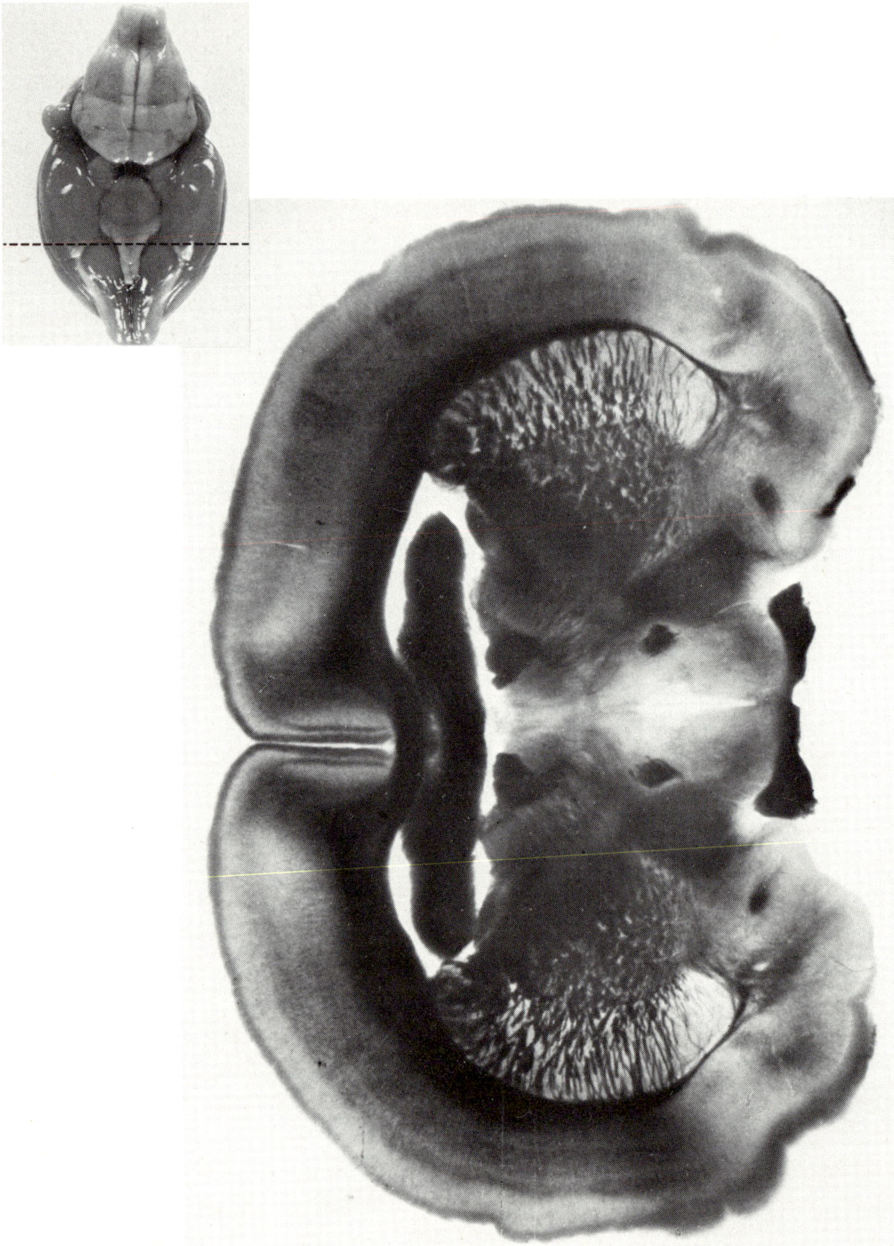

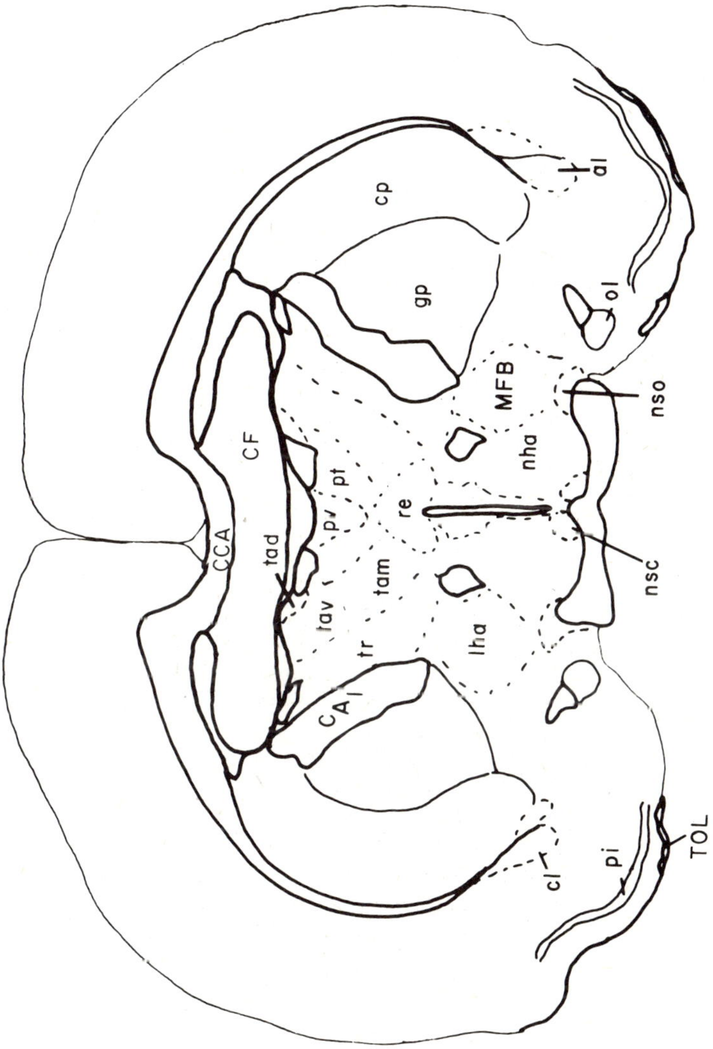
cp
al
gp
ol
MFB
nso
nha
CF
pt
pl
re
CCA
tad
tam
tav
tr
lha
nsc
CA
TOL
pi
cl

Figure 14

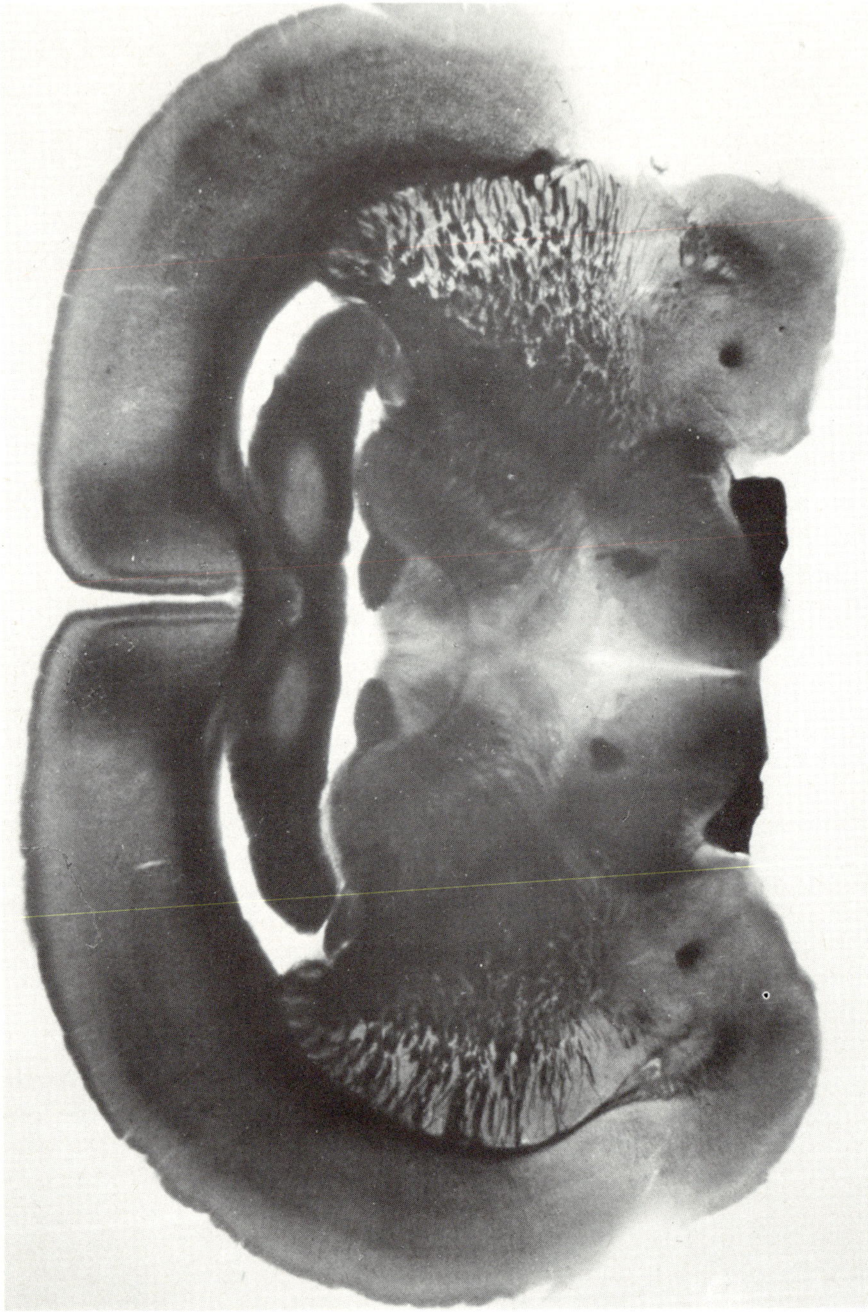

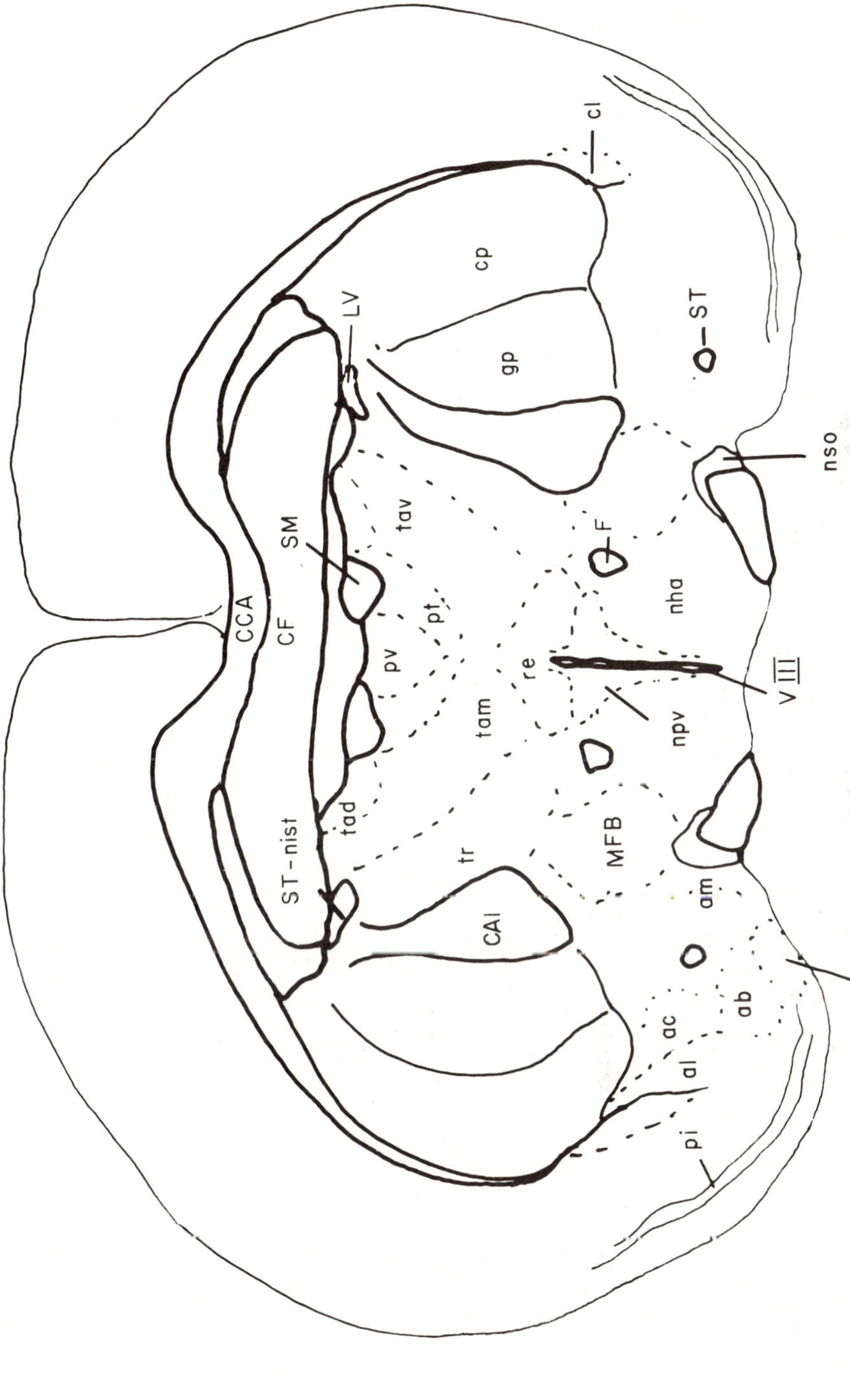
cl
cp
gp
LV
ST
nso
F
tav
SM
nha
pt
CCA
CF
pv
re
V III
tam
npv
tad
ST-nist
tr
MFB
am
CAI
aco
ab
ac
al
pi

Figure 15

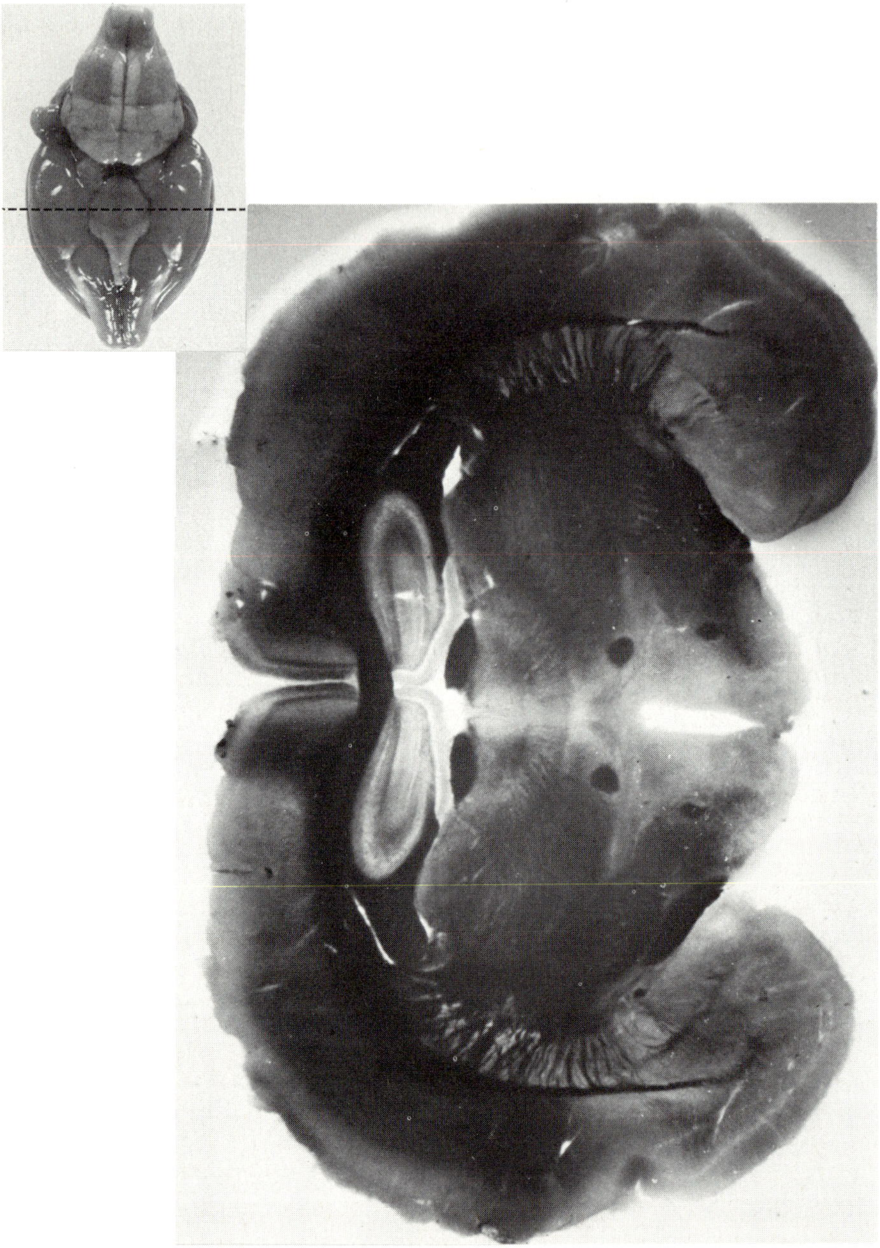

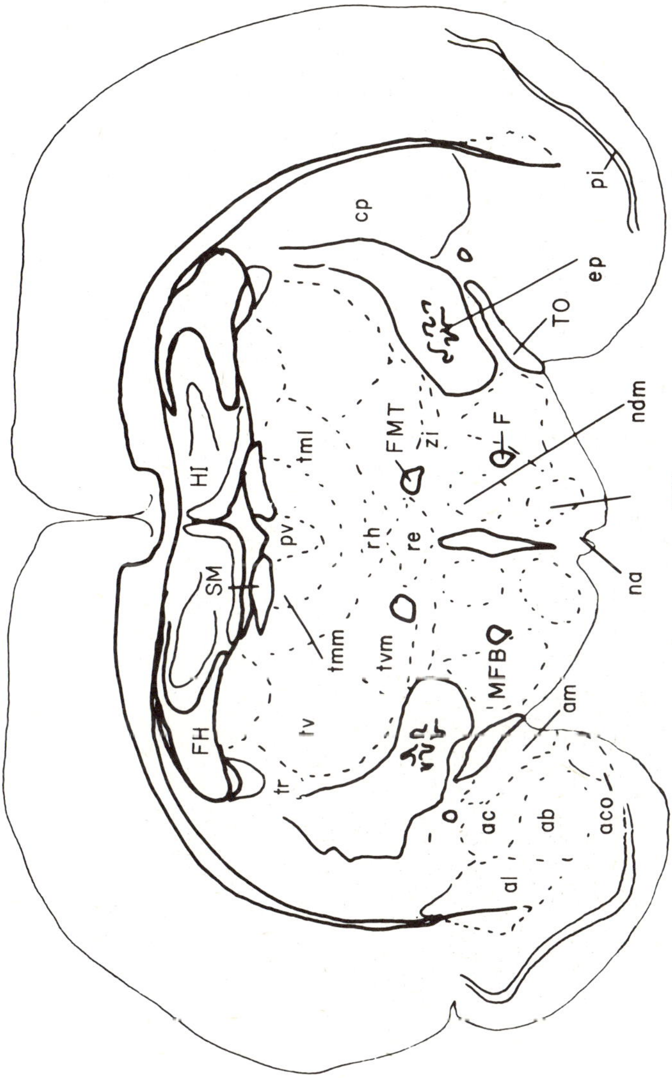
cp
pi
ep
TO
ndm
nvm
na
F
FMT
zi
tml
HI
pv
rh
re
SM
tmm
tvm
MFB
am
FH
tv
tr
ac
ab
aco
al

Figure 16

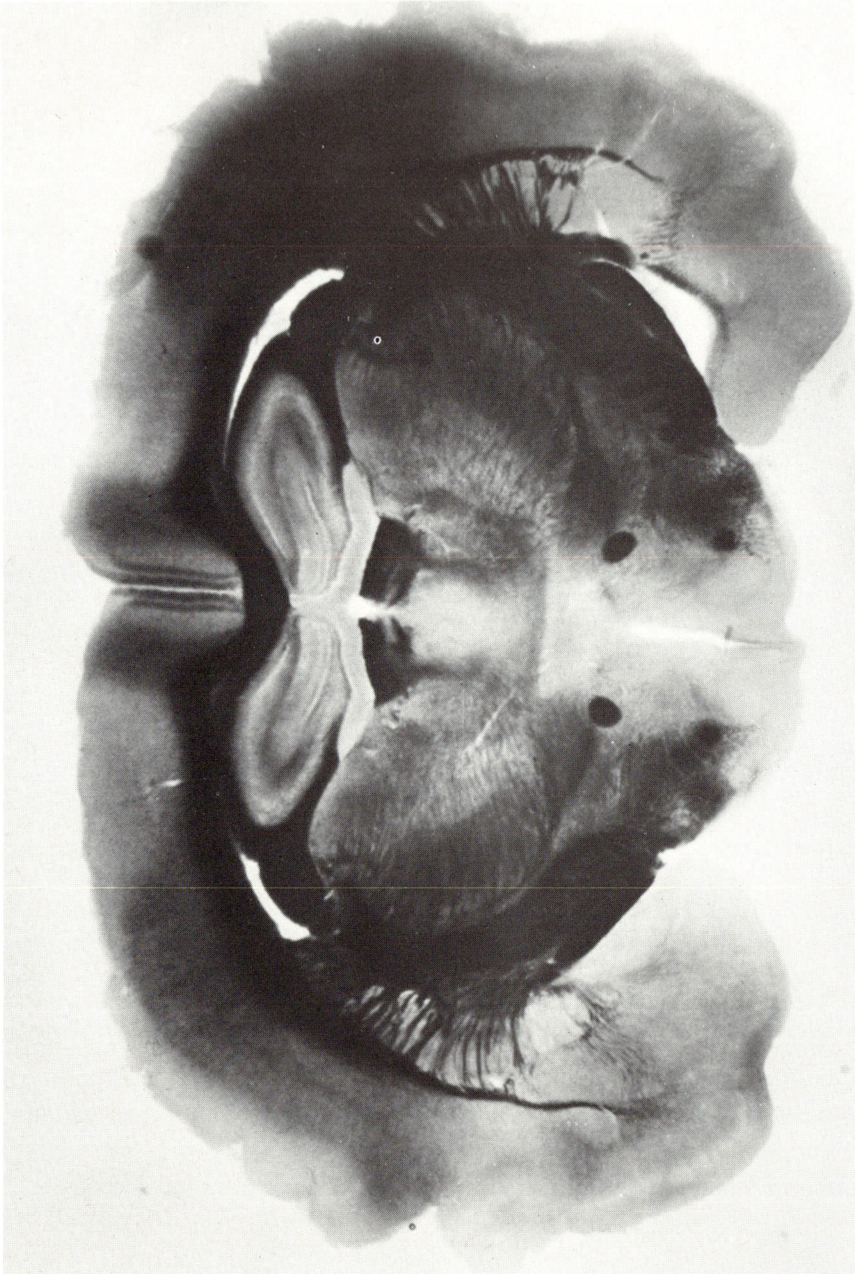

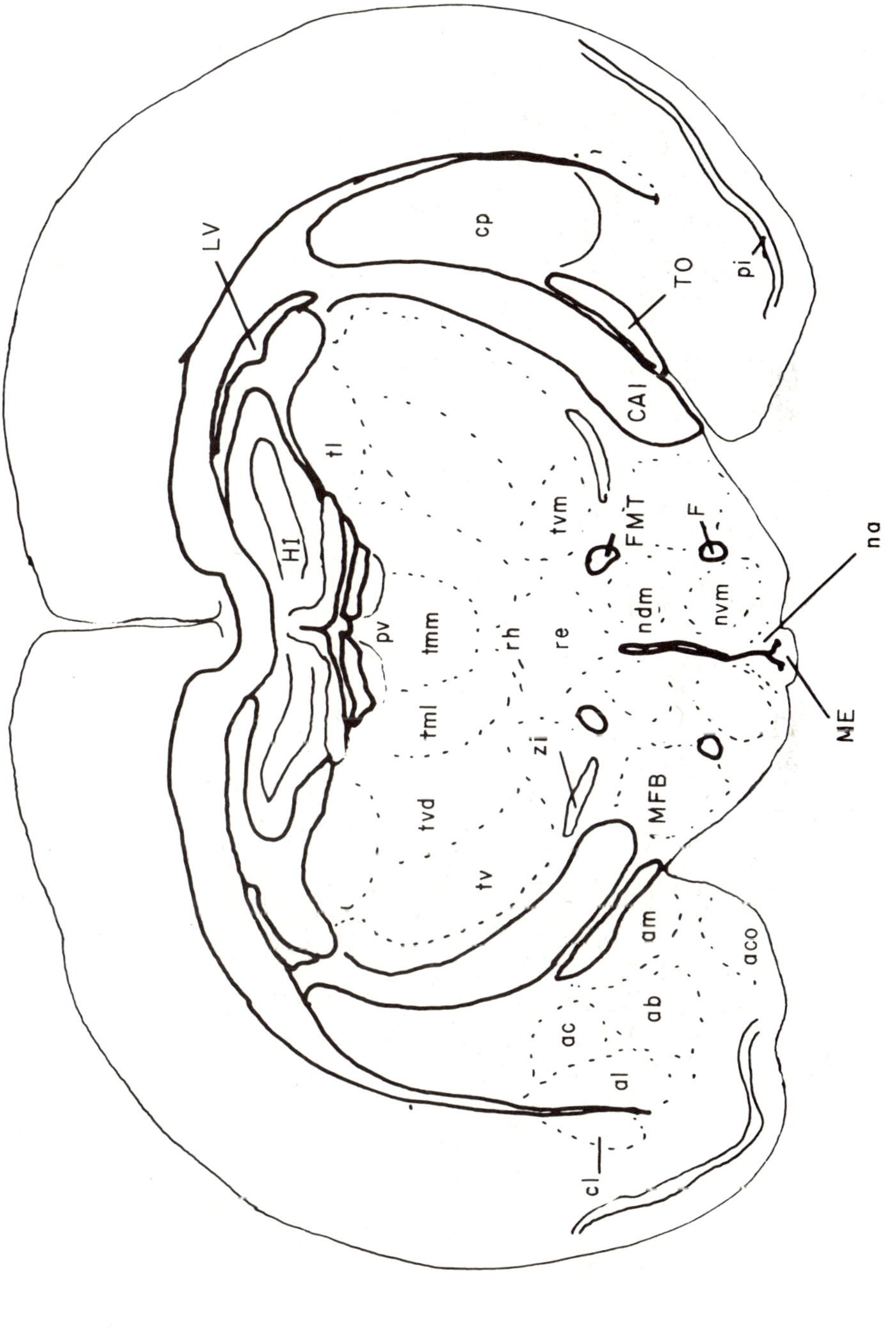
LV
cp
TO
pi
CAI
tl
tvm
FMT
F
HI
na
pv
tmm
rh
re
ndm
nvm
ME
tml
zi
tvd
MFB
tv
am
aco
ab
ac
al
cl

Figure 17

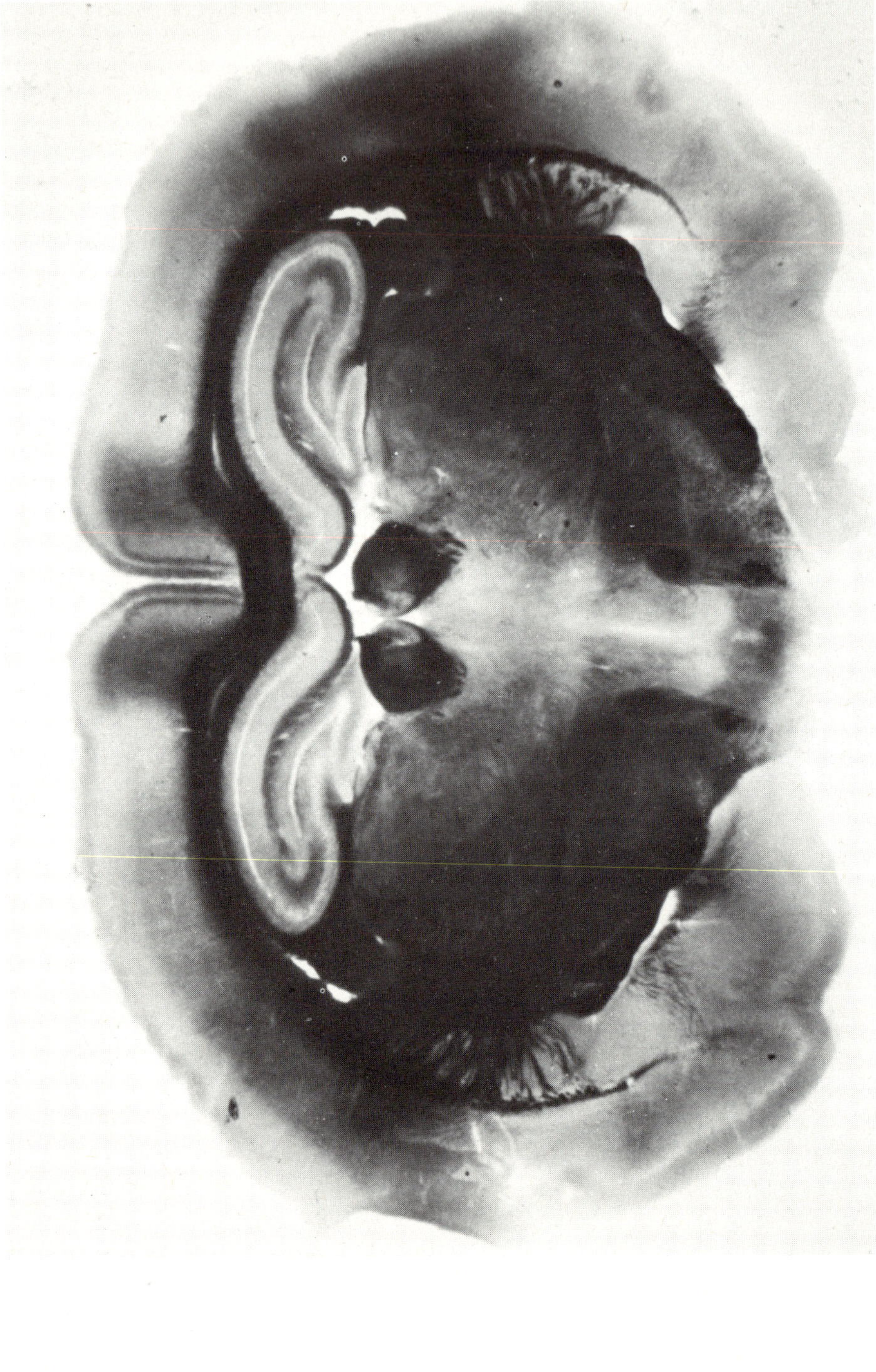

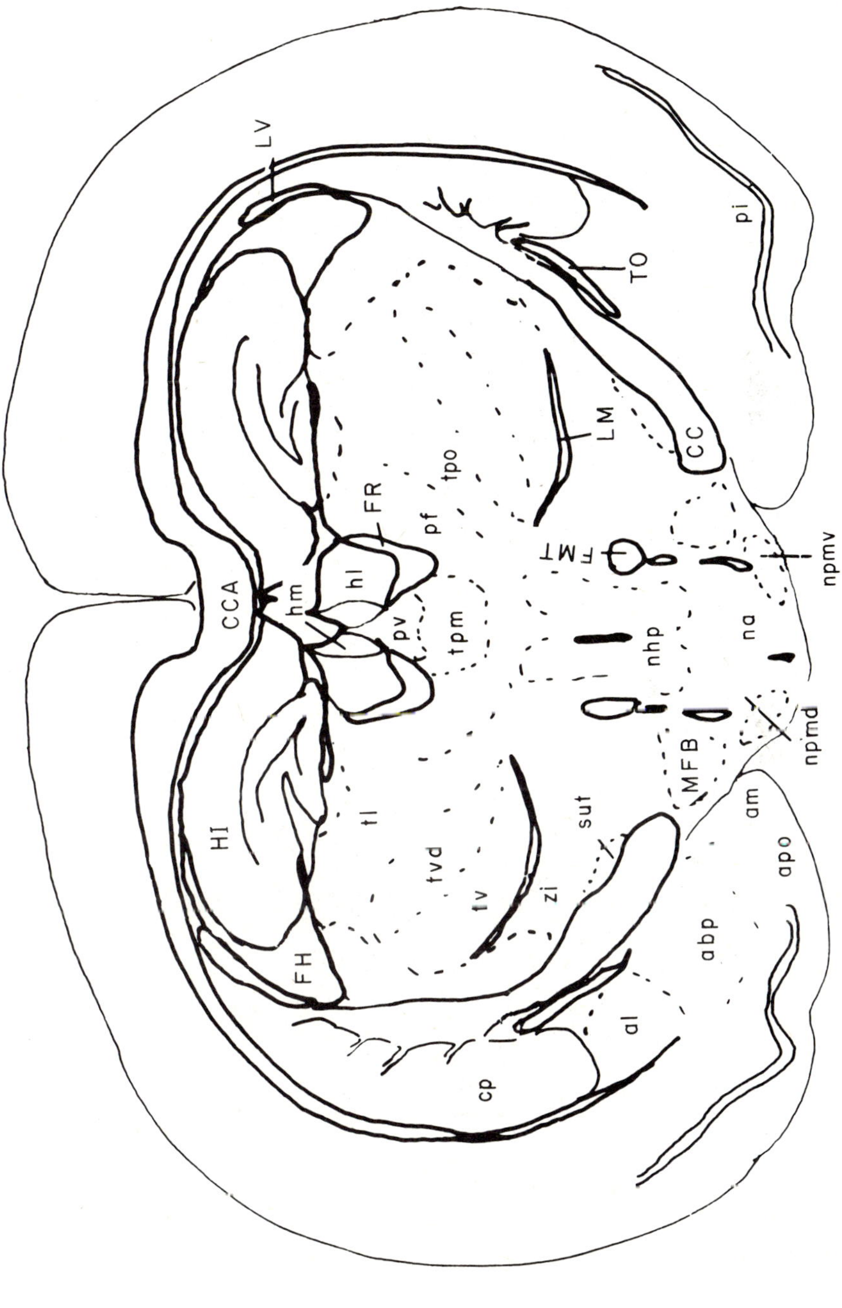
LV
pi
TO
LM
CC
FR
tpo
pf
FMT
npmv
hl
hm
CCA
pv
tpm
nhp
na
npmd
MFB
am
apo
sut
tl
tvd
tv
zi
HI
FH
abp
al
cp

Figure 18

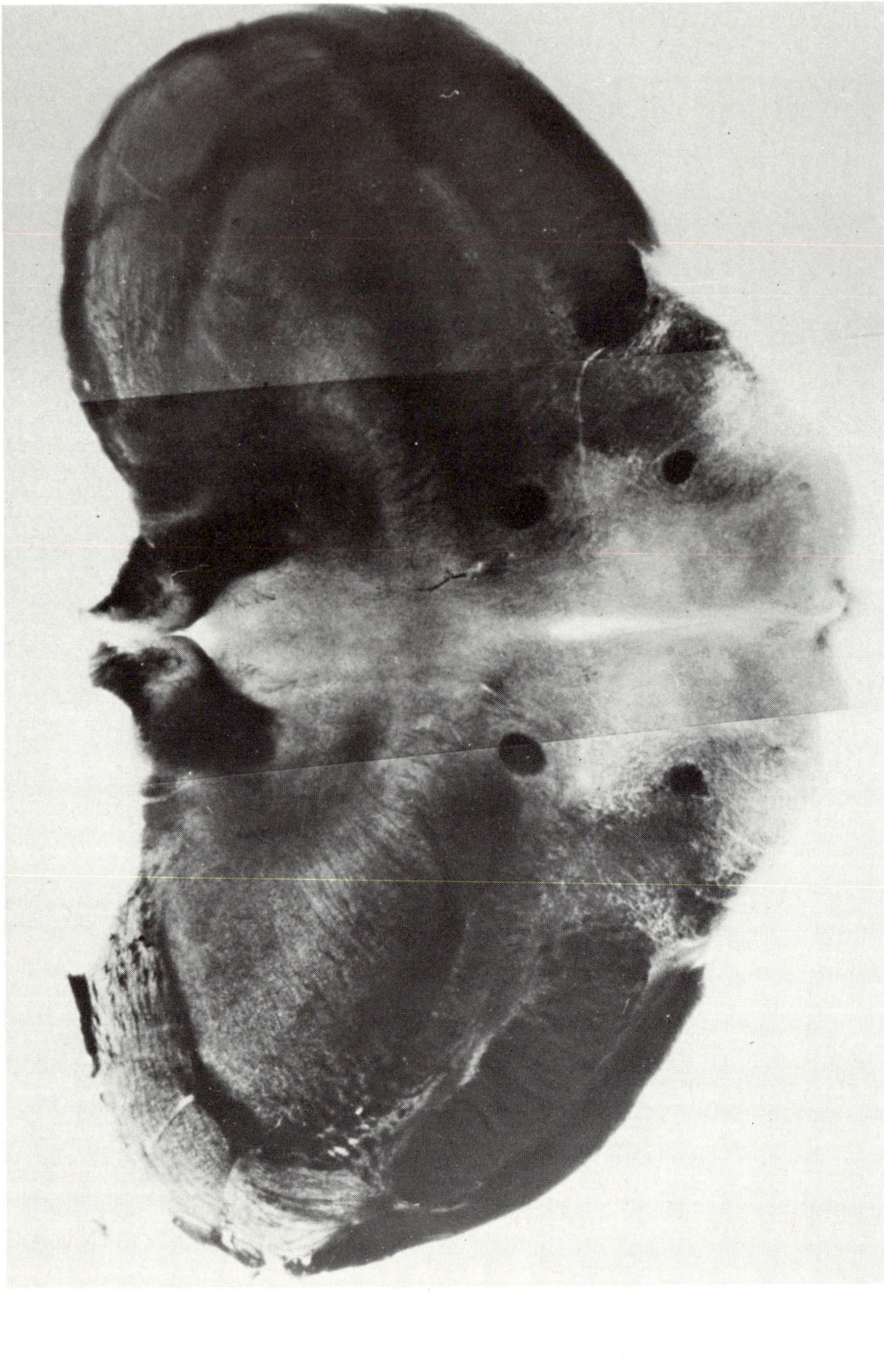

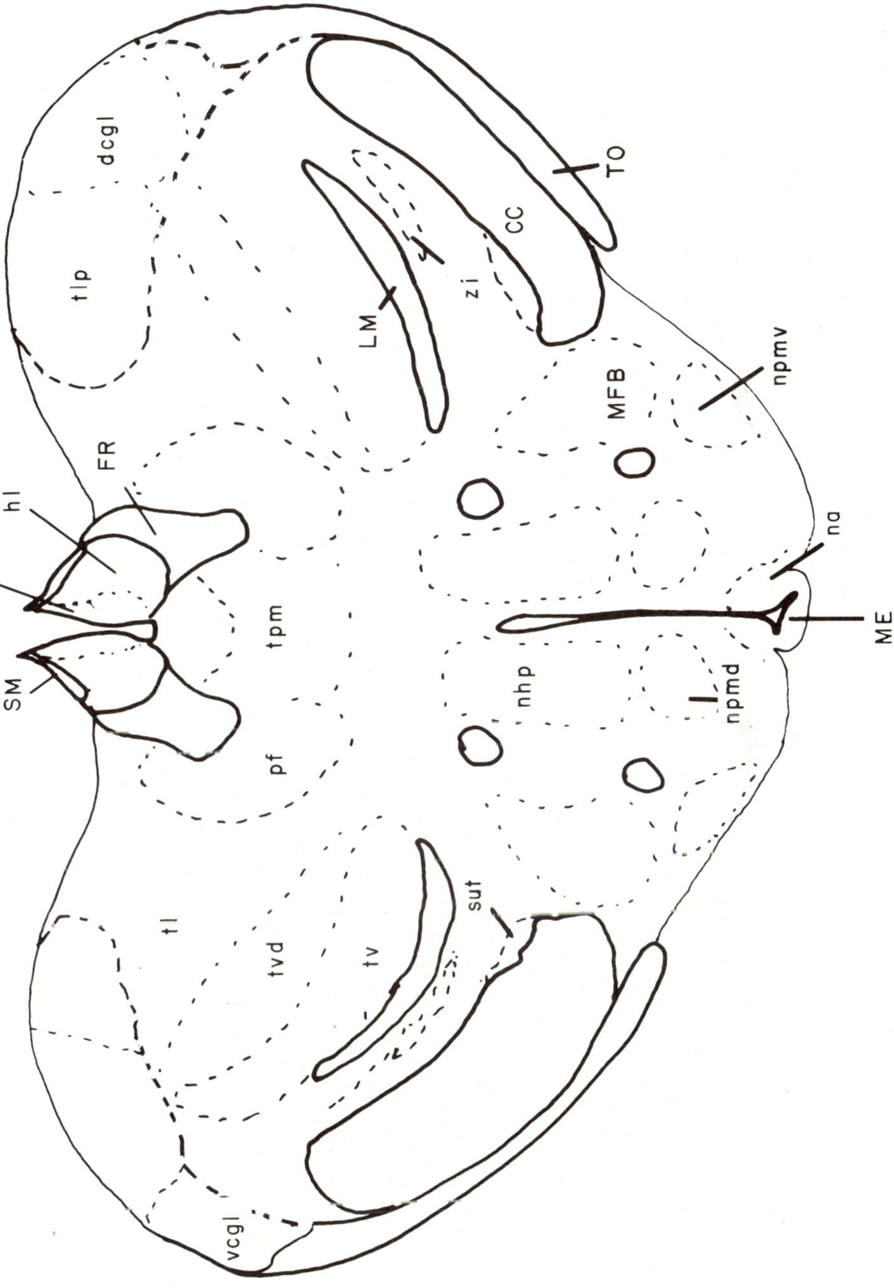
dcgl
tlp
FR
hl
hm
SM
tpm
pf
tl
tvd
tv
sut
vcgl
TO
CC
zi
LM
MFB
npmv
na
ME
nhp
npmd

Figure 19

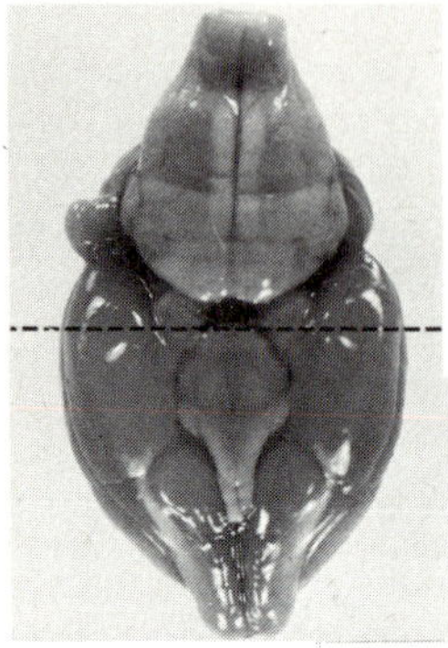

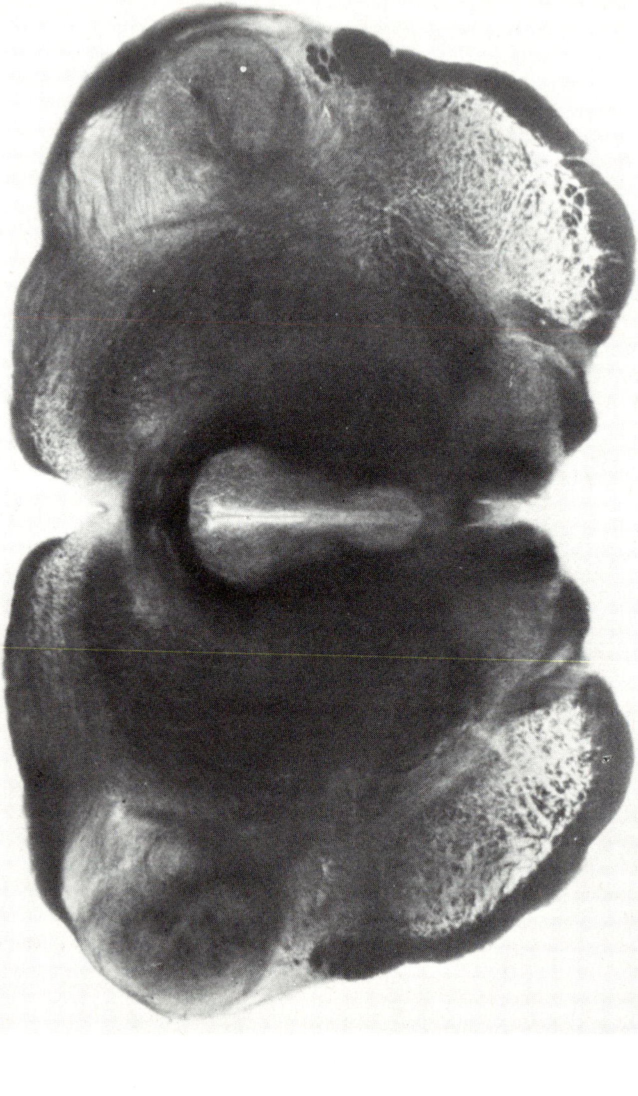

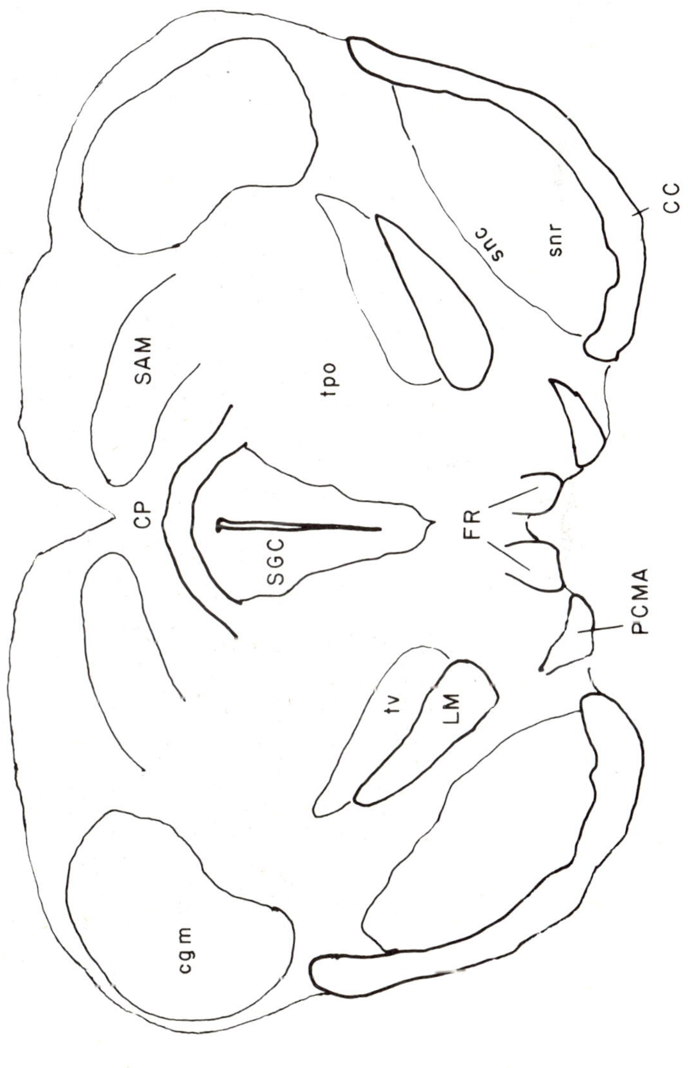

Figure 20

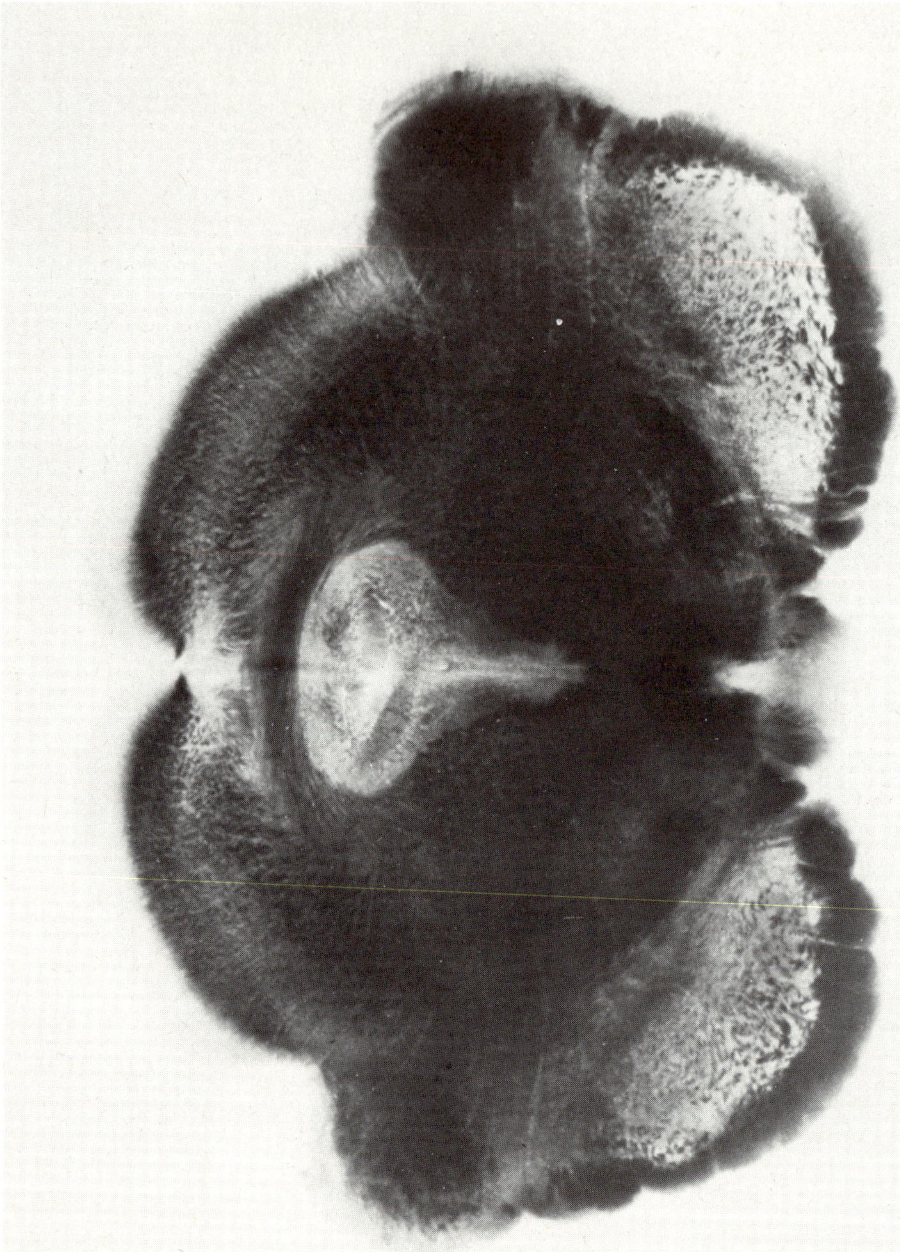

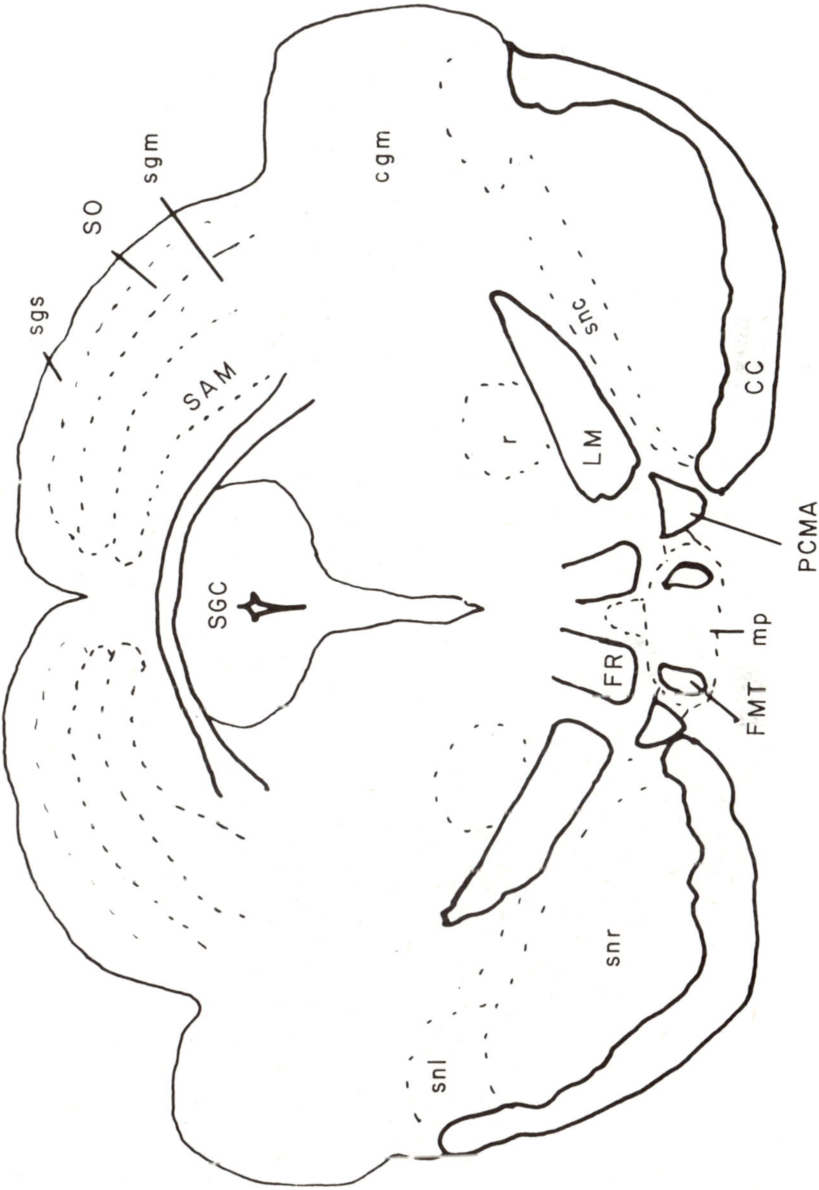

Figure 21

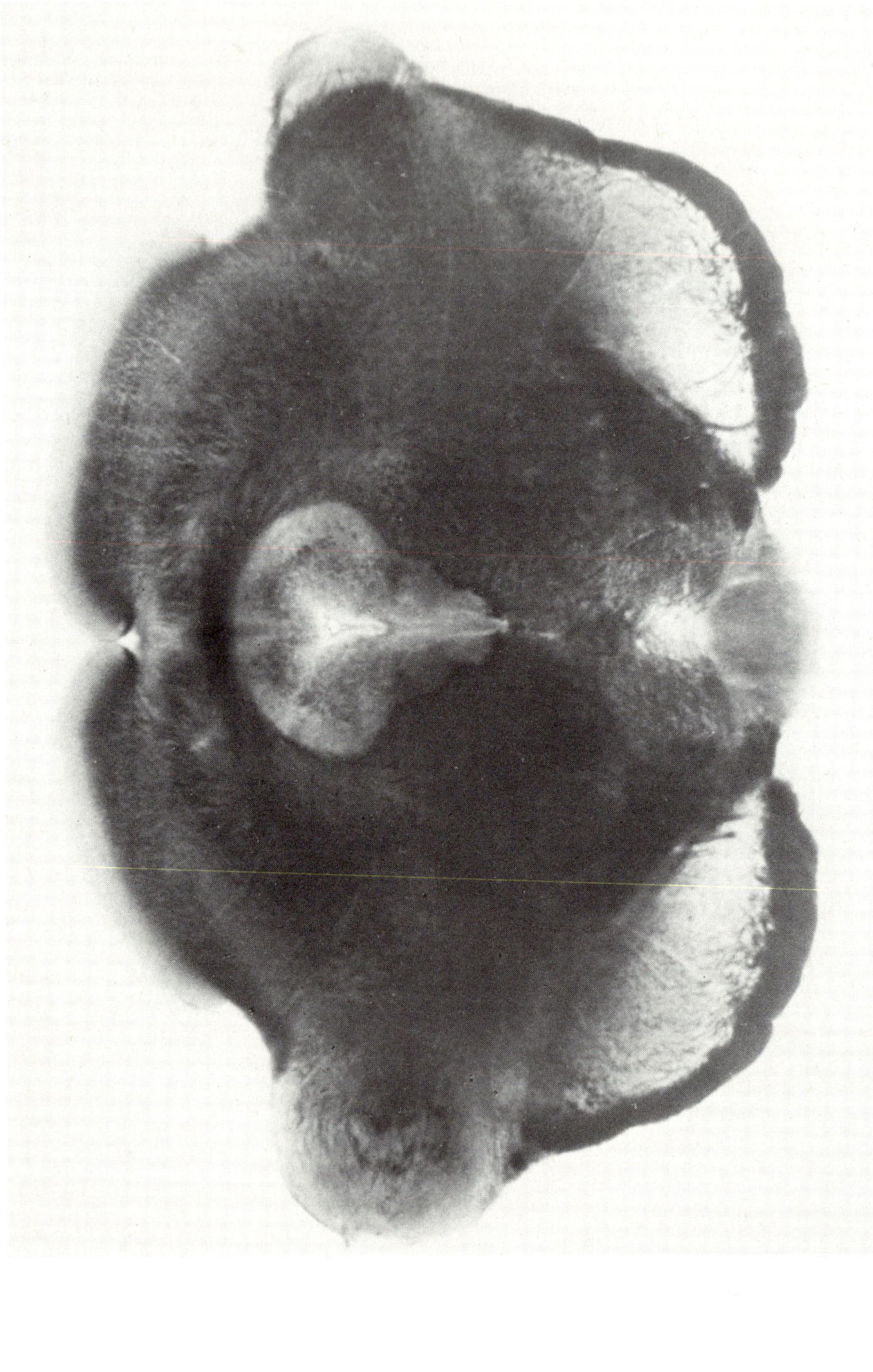

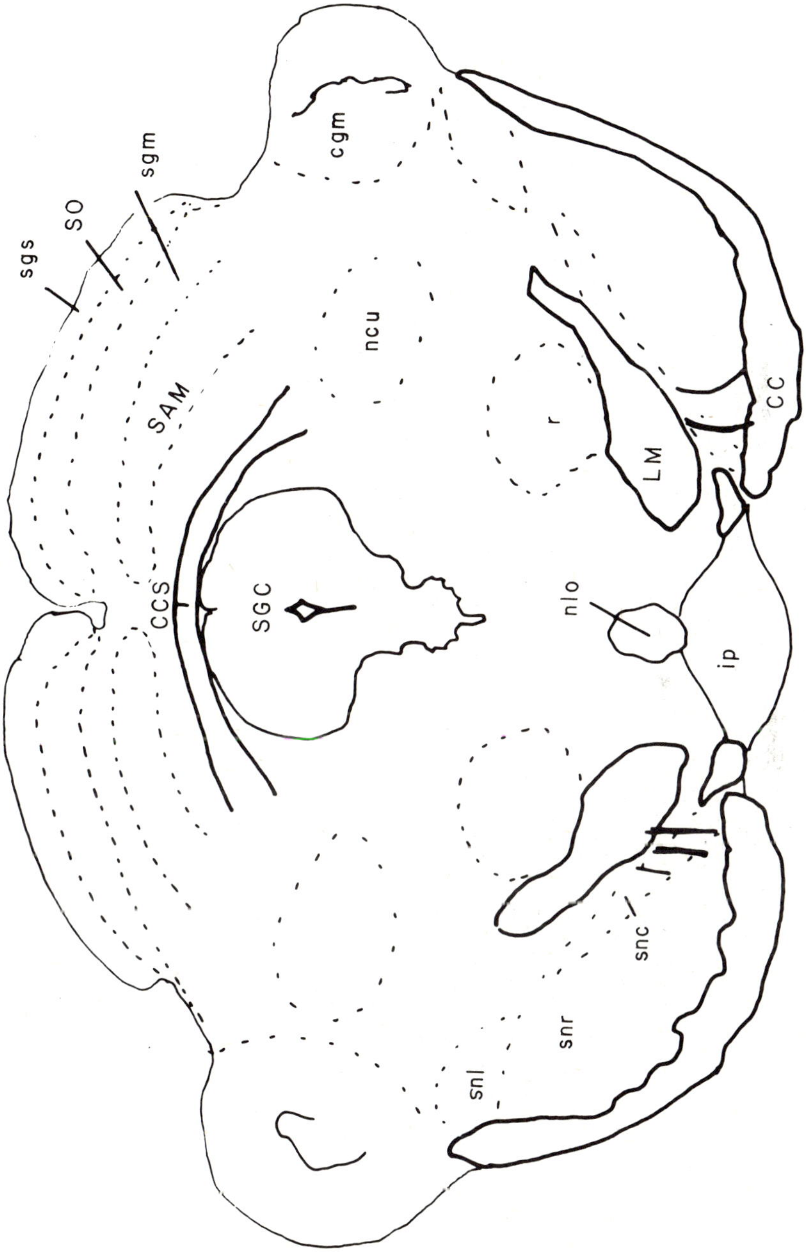

Figure 22

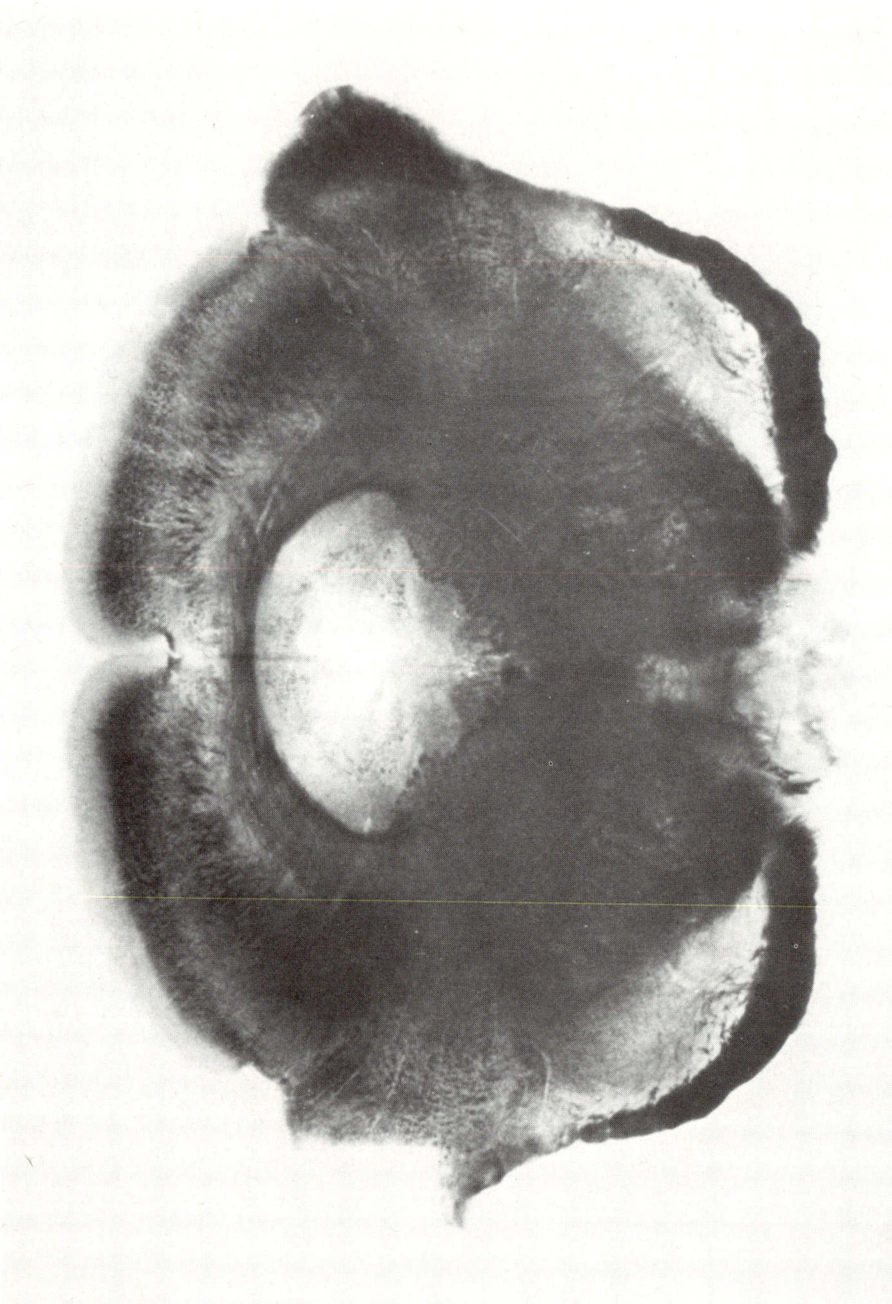

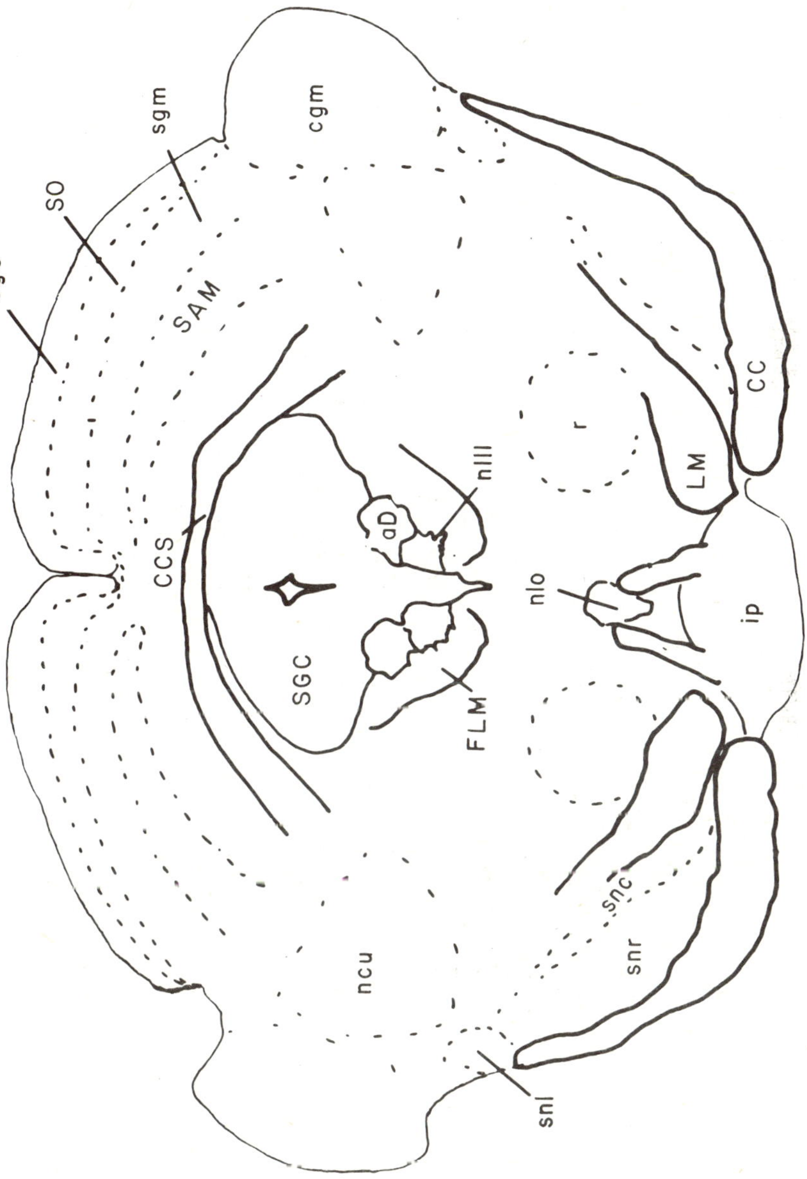
sgs
SO
sgm
cgm
SAM
CC
r
LM
nIII
aD
CCS
nlo
ip
SGC
FLM
snc
snr
ncu
snl

Figure 23

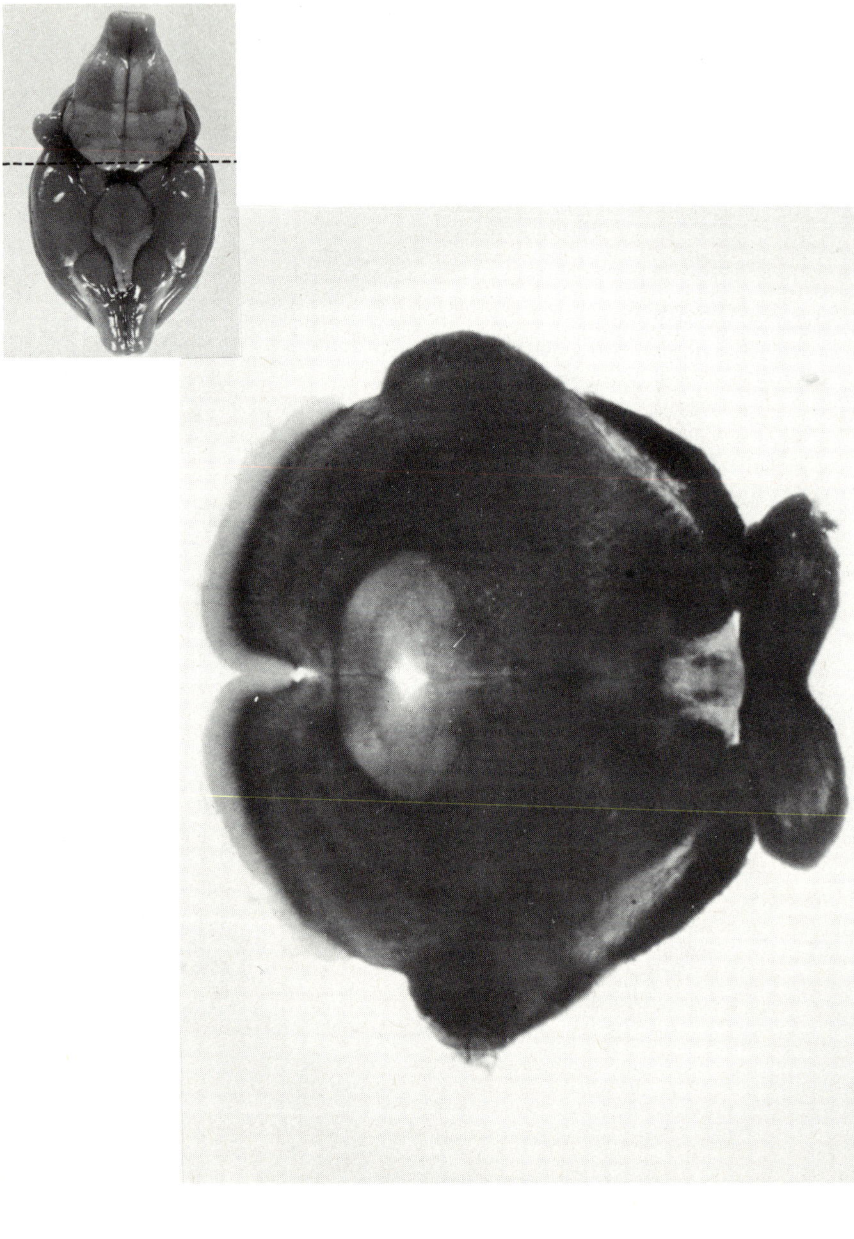

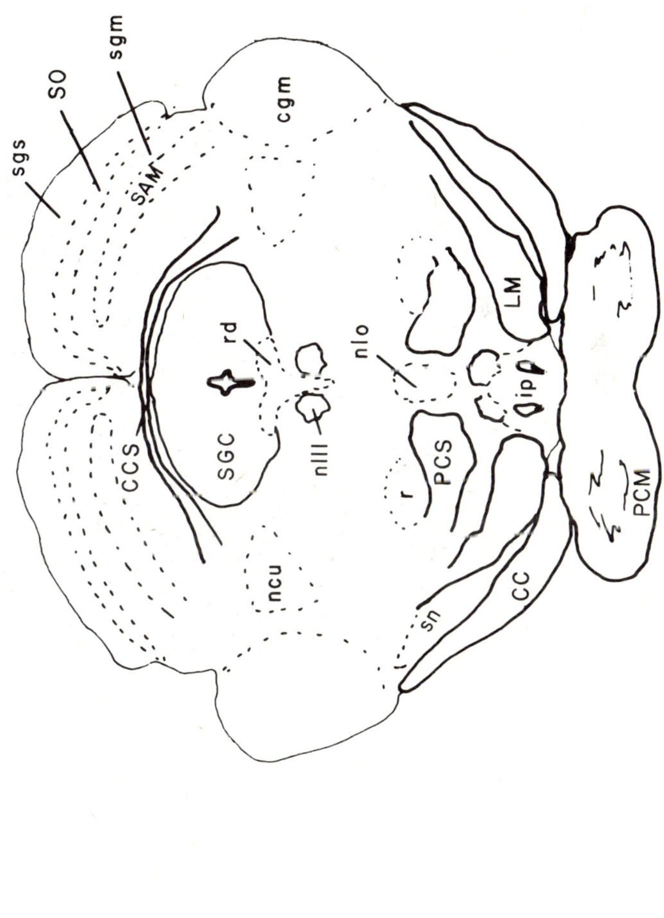

Figure 24

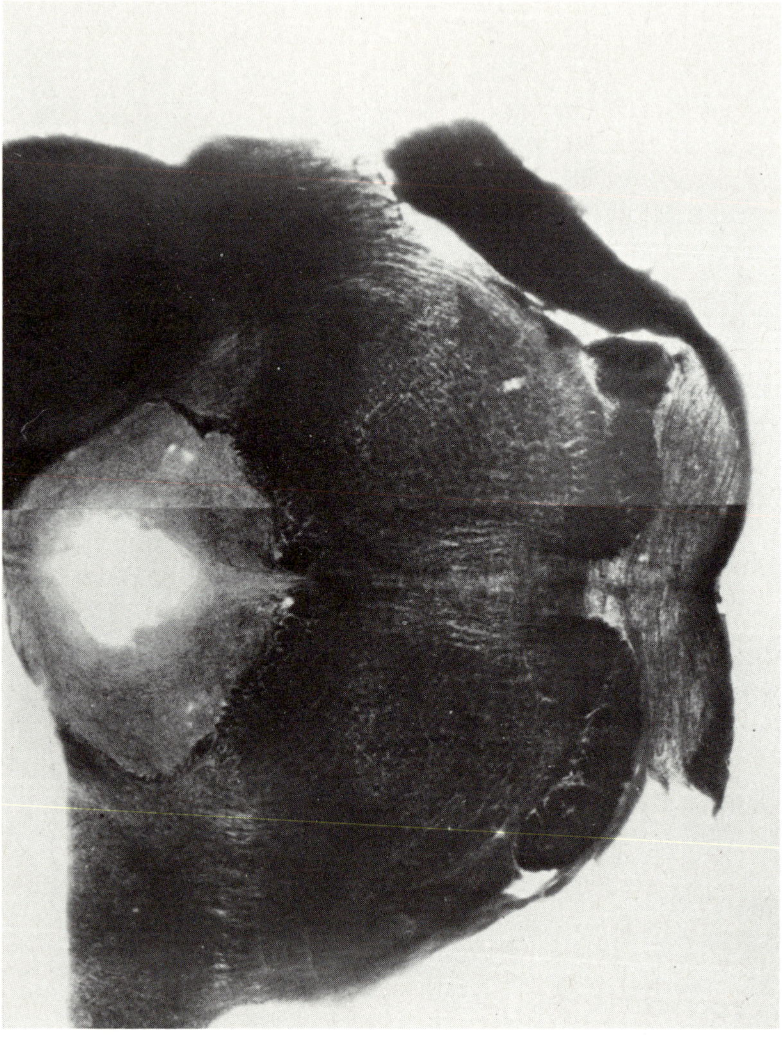

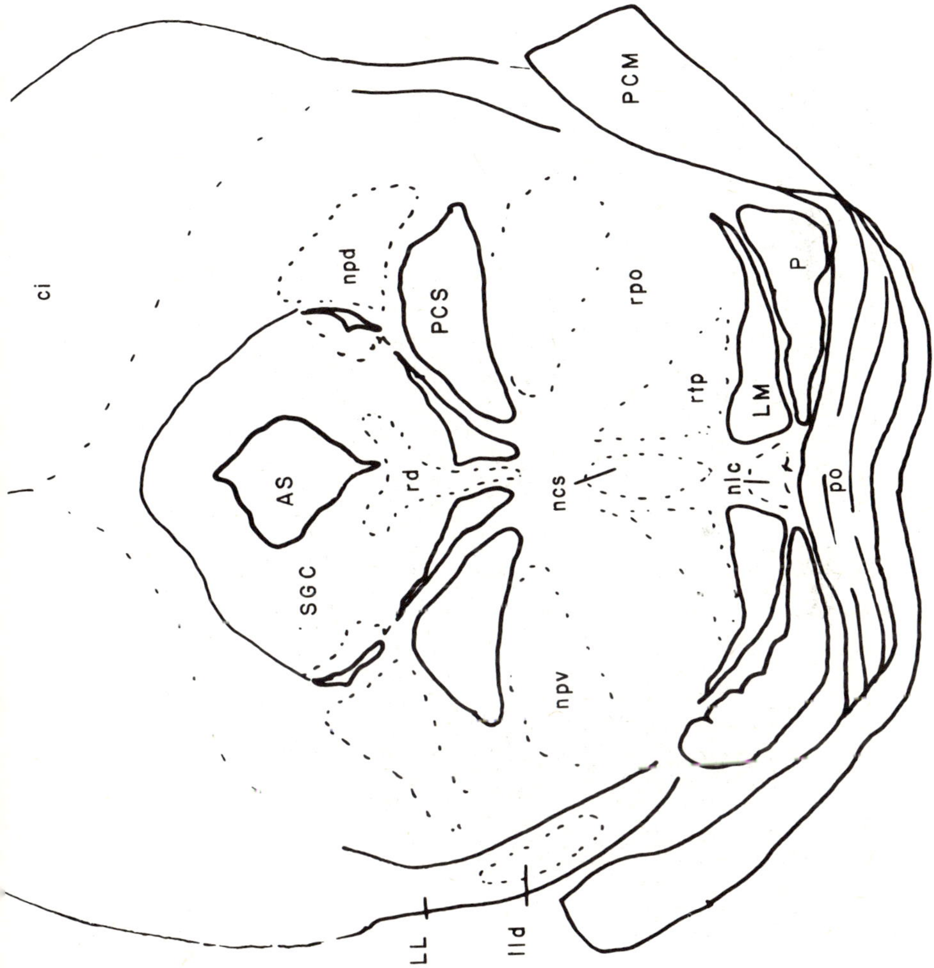

Figure 25

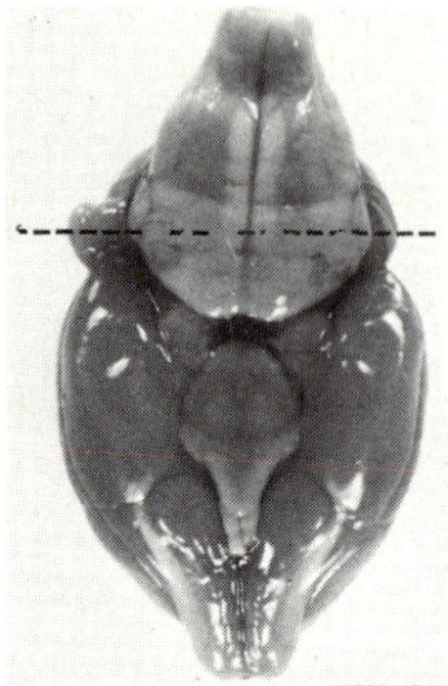

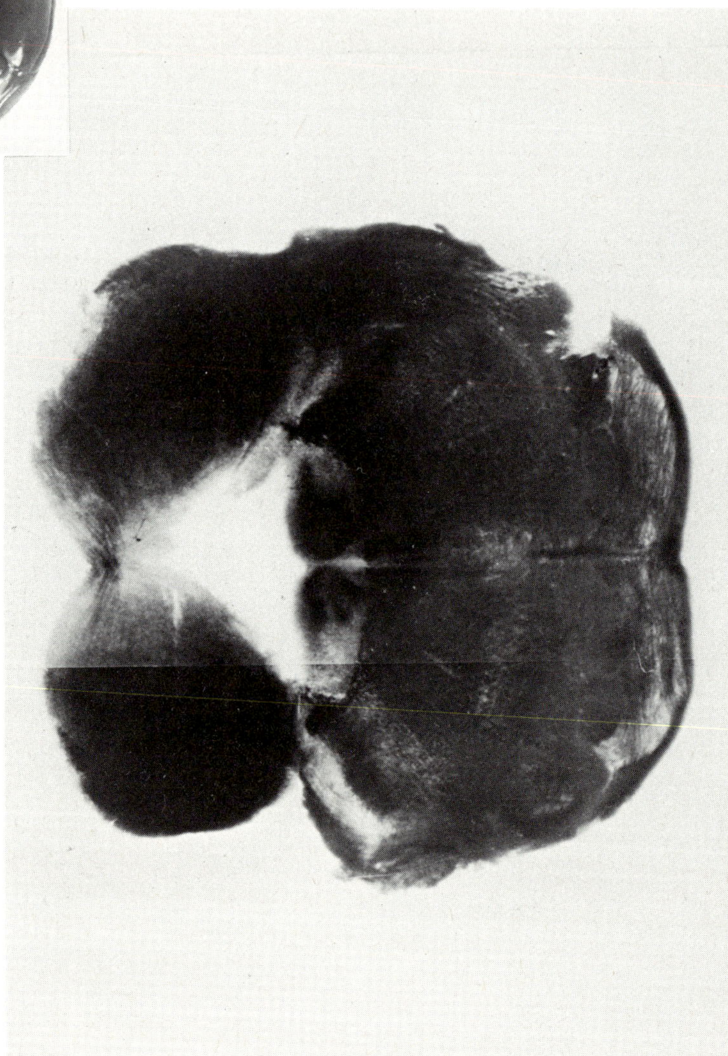

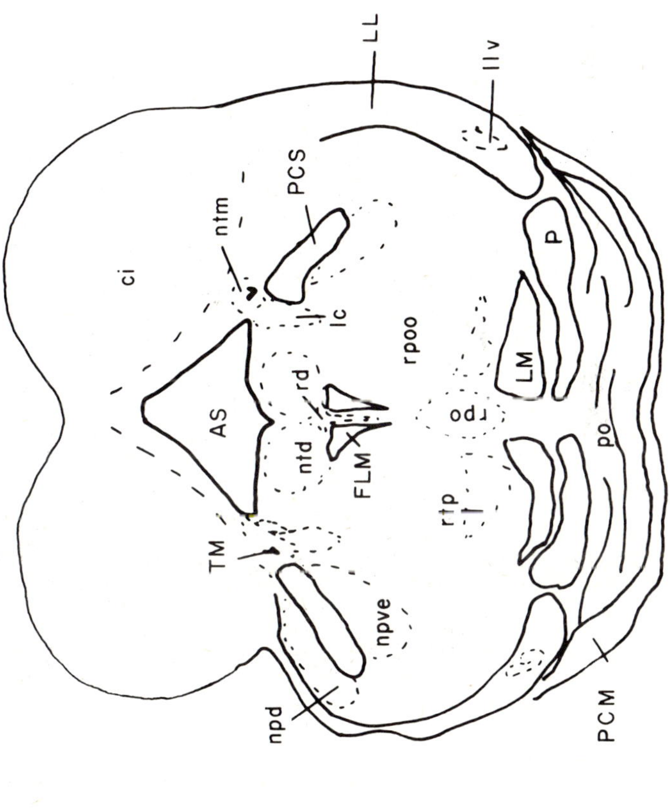
LL
llv
PCS
ntm
ci
lc
rpoo
AS
rd
ntd
FLM
rpo
LM
P
po
rtp
TM
npve
npd
PCM

Figure 26

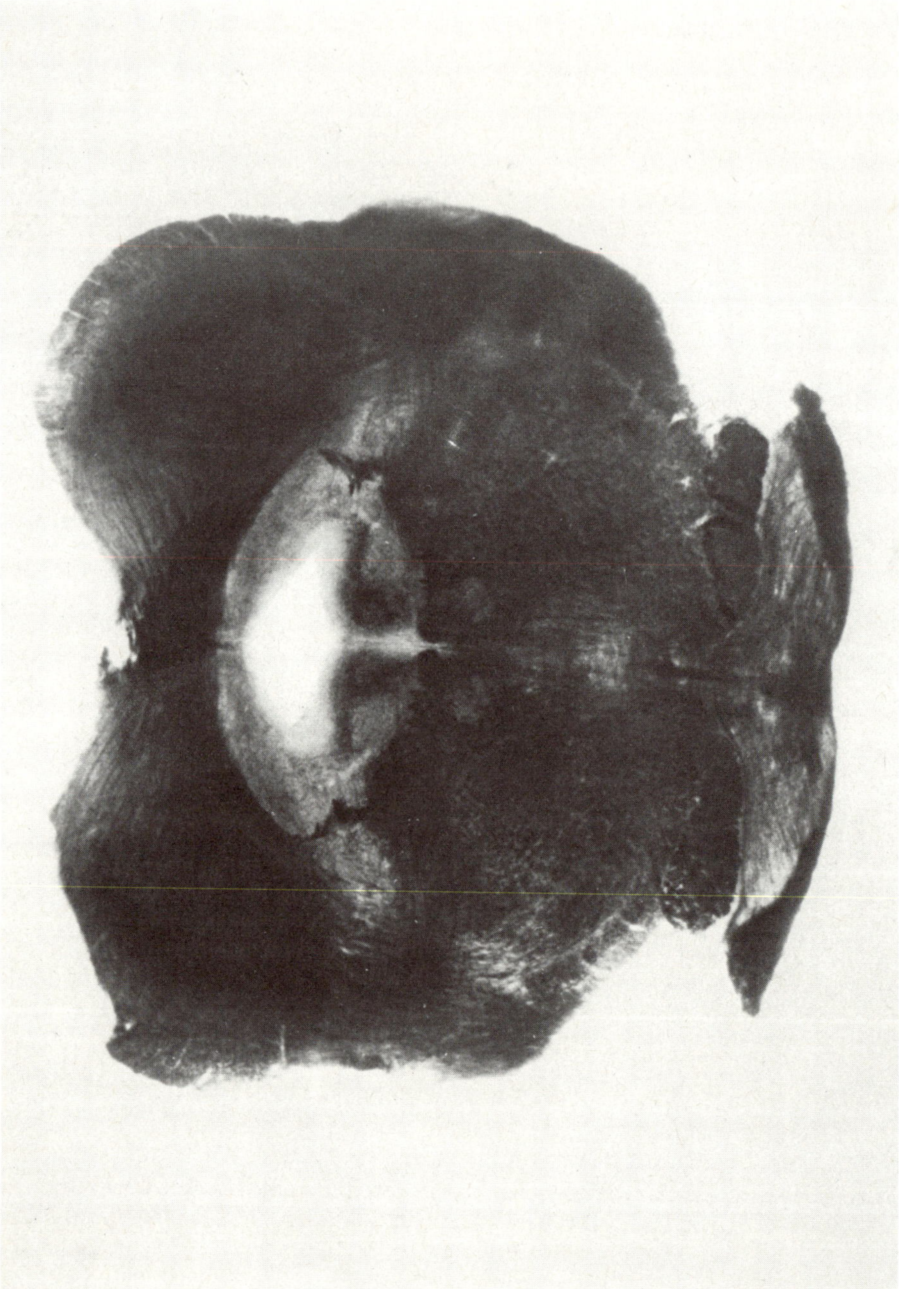

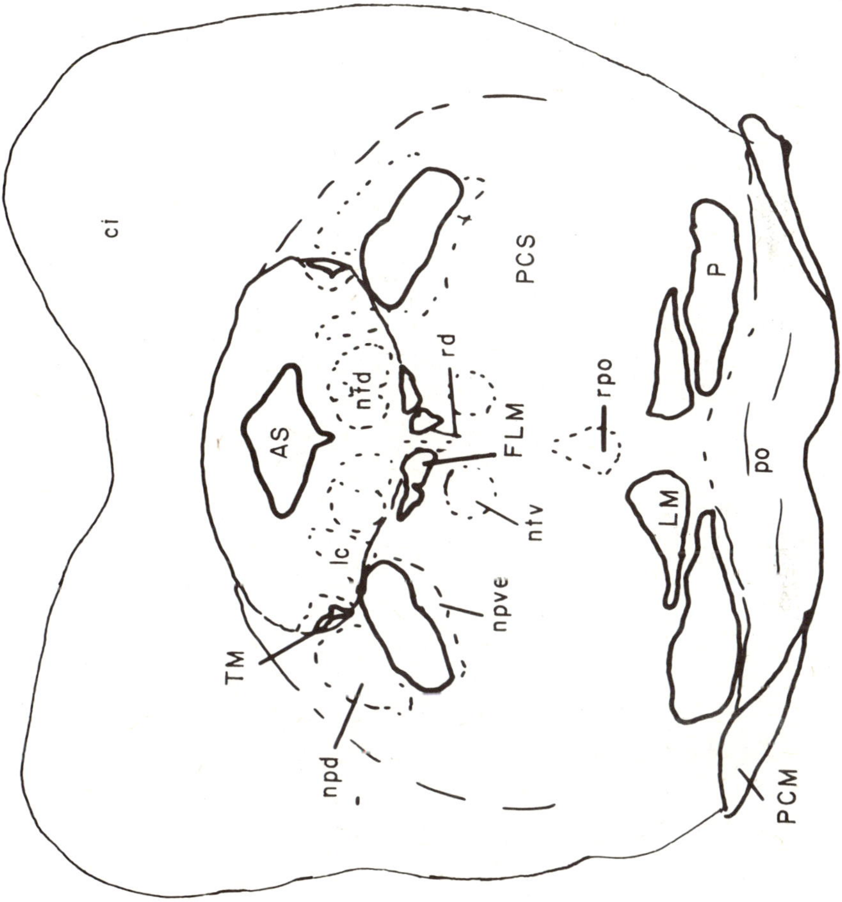

Figure 27

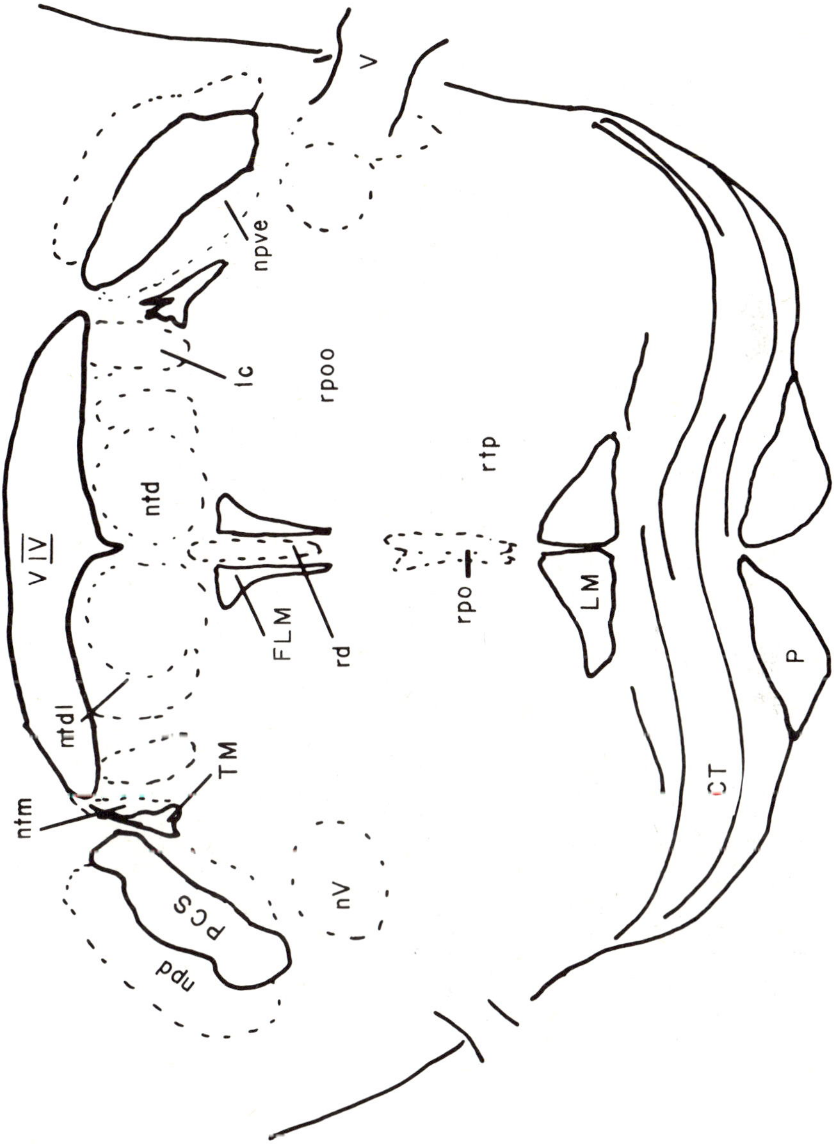
V
npve
lc
rpoo
rtp
ntd
V IV
FLM
rd
rpo
LM
P
CT
ntdl
TM
ntm
nV
PCS
npd

Figure 28

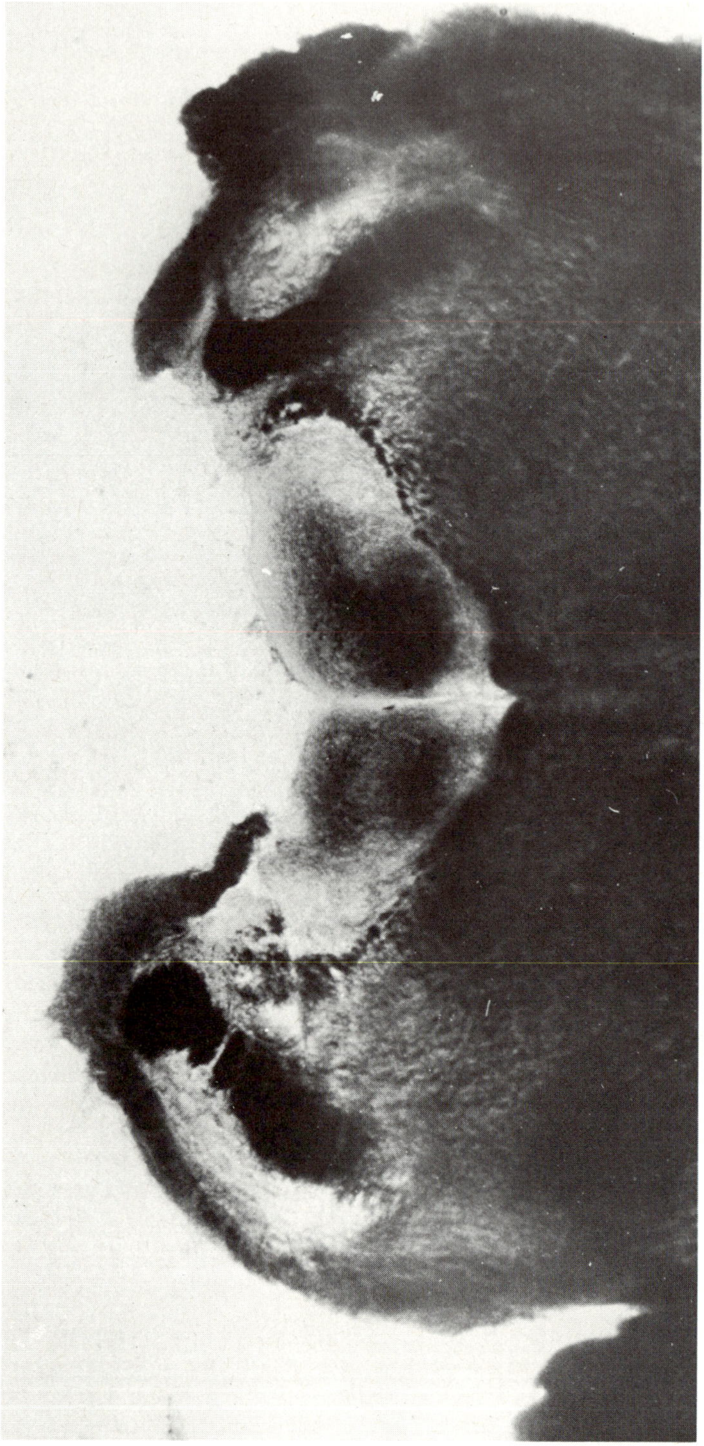

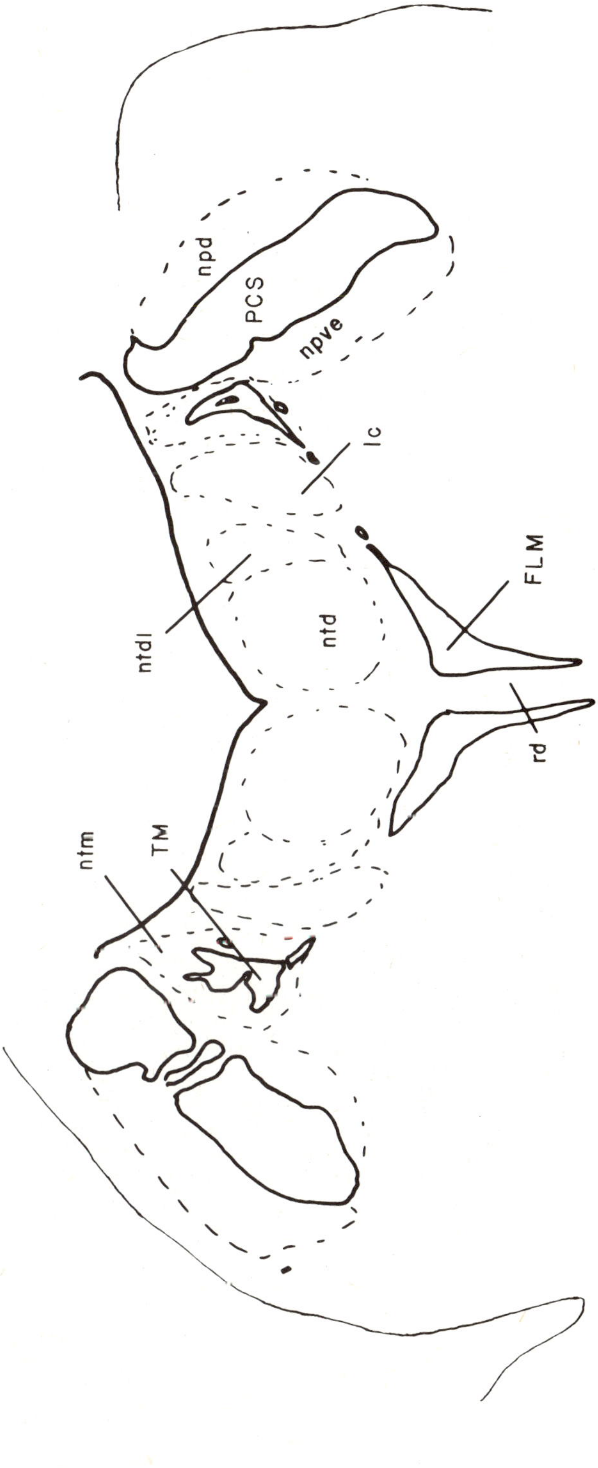

Figure 29

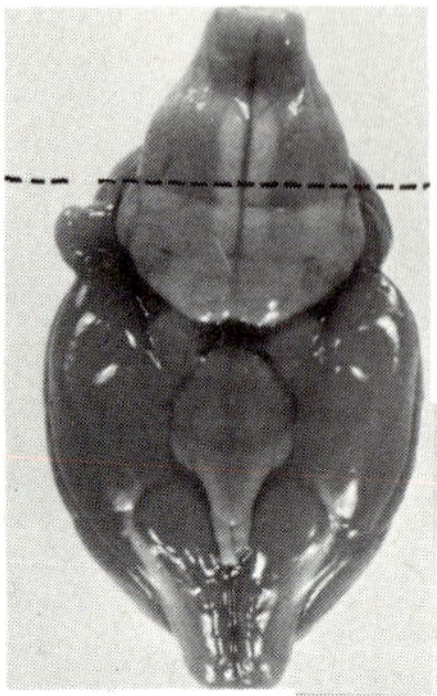

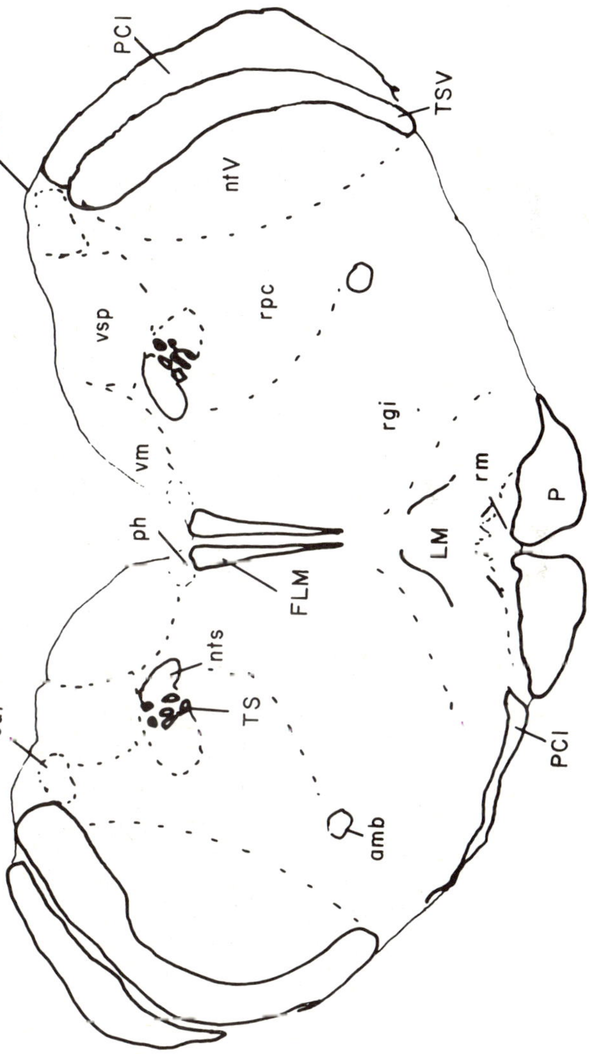

Figure 30

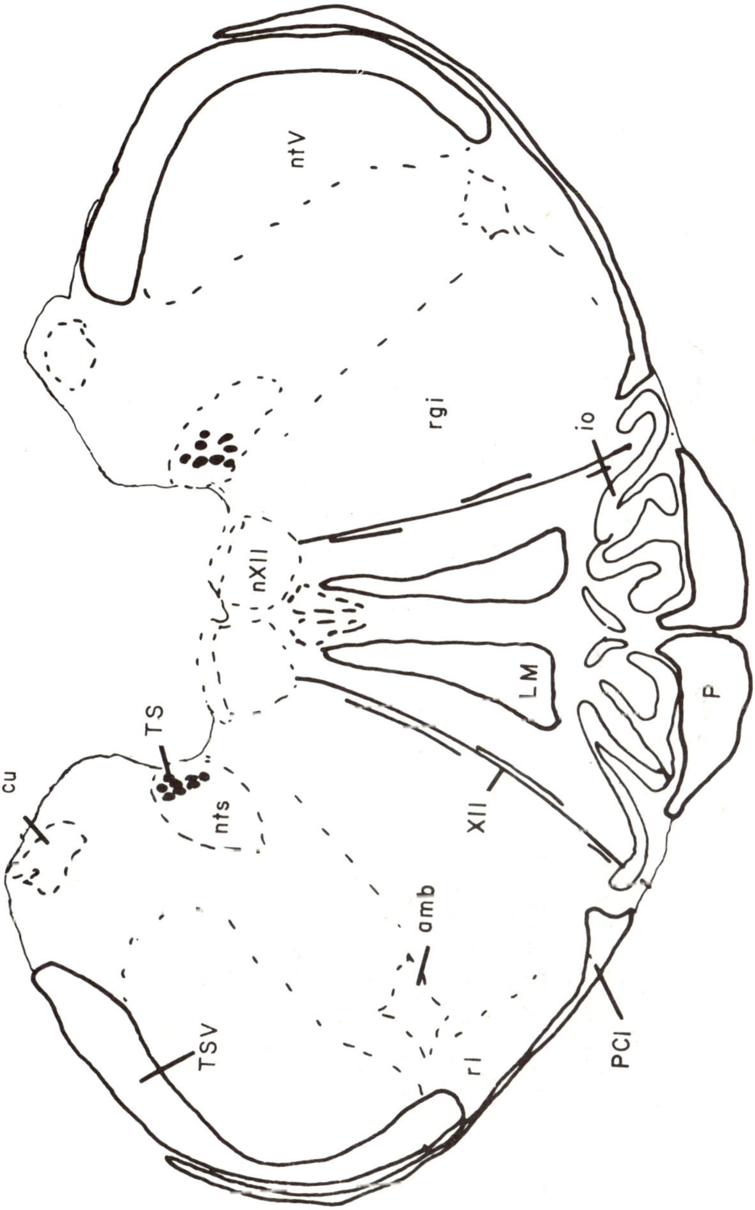

Figure 31

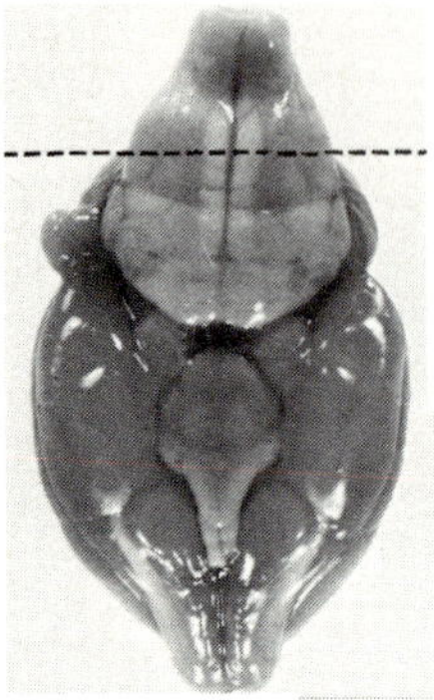

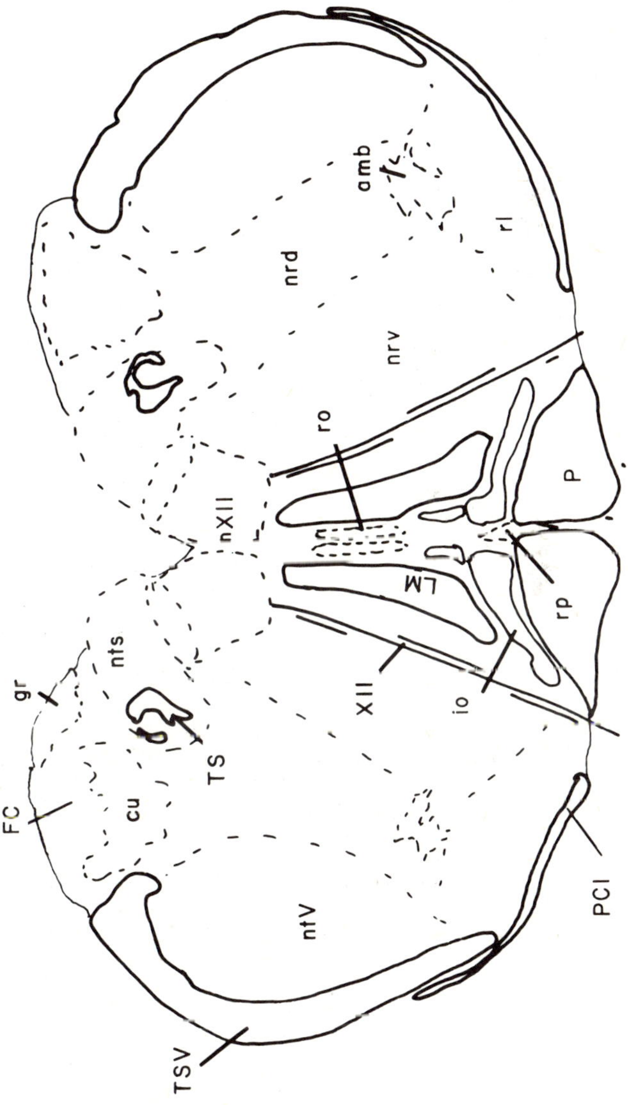

Figure 32

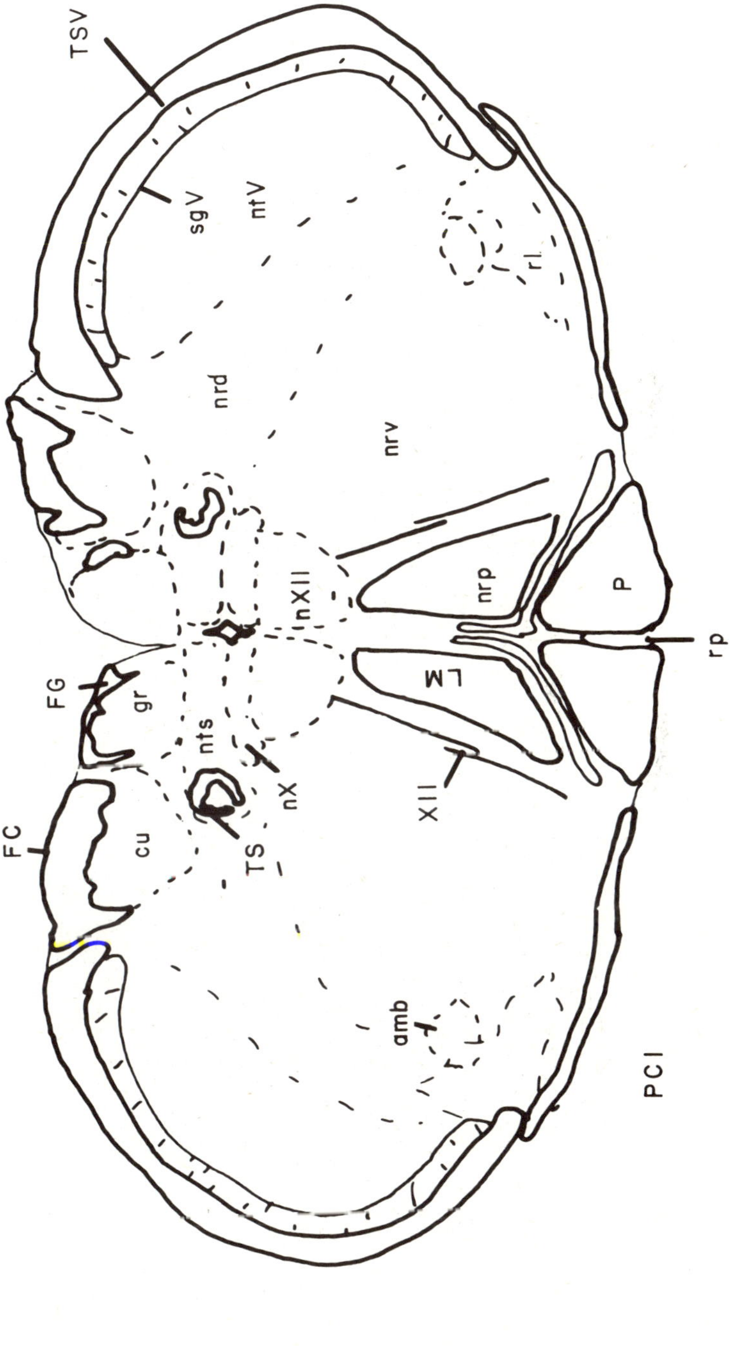
TSV
sgV
ntV
rl
nrd
nrv
nXII
nrp
P
rp
LM
XII
FG
gr
nts
nX
FC
cu
TS
amb
PCI

Figure 33

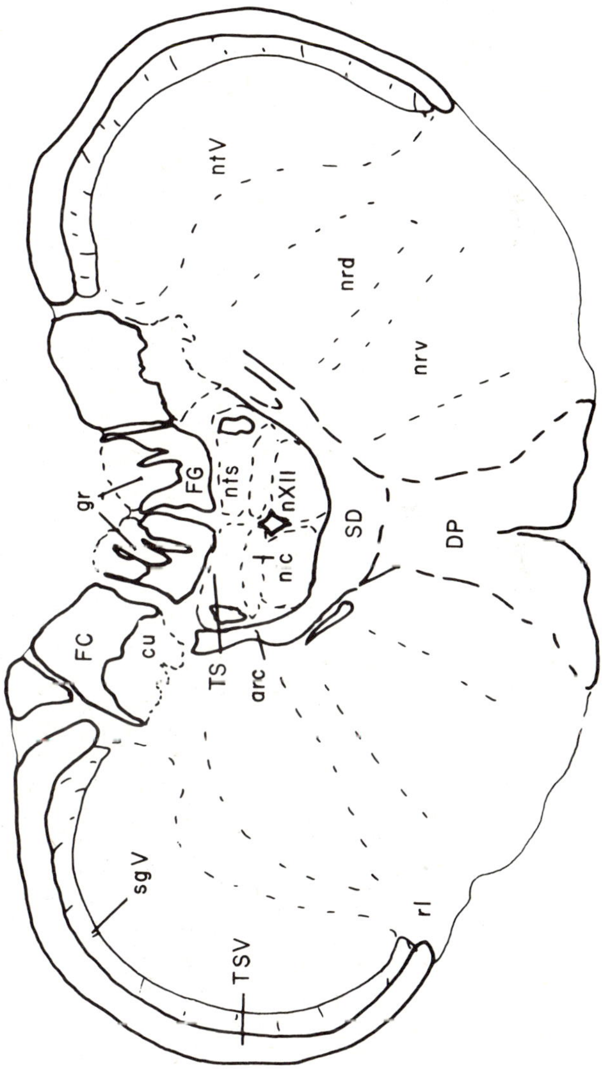

Figure 34

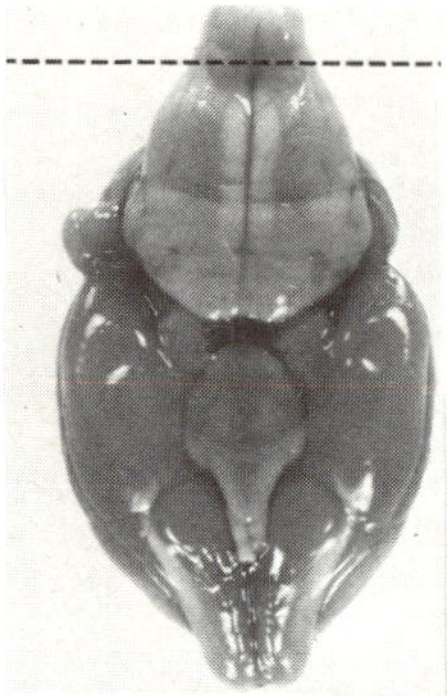

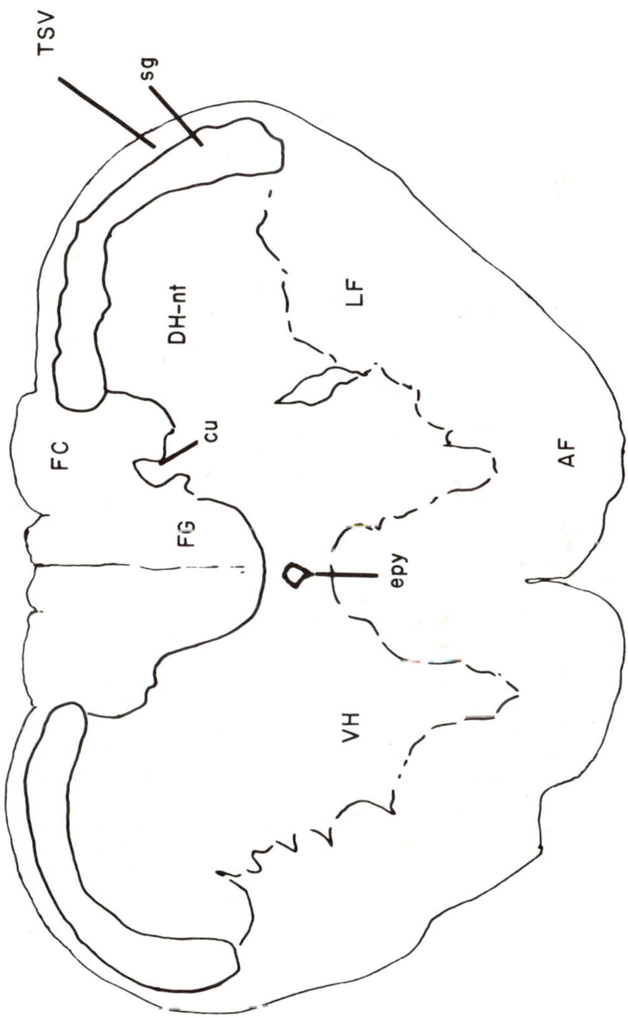

Figure 35

FP4 22 DIN/125 ASA). The use of higher magnification allows even better resolution of nuclear structures up to single neurons of motor nuclei.

The slices (all obtained with the McIlwain tissue chopper) represented in these figures correspond to sections of 250 μm for the telencephalic–diencephalic–mesencephalic levels while thinner slices are needed for the visualization of nuclear structures in rhomboencephalic levels (pons (125 μm) and medulla oblongata (150 μm)). This is due to the very high concentration of myelinated structures in these lower levels of the brain stem. At any level the heavily myelinated commissural fibre systems such as the corpus callosum (CCA) or the anterior commissure (CA) as well as ascending and descending tracts, notably the fasciculus gracilis (fg) and cuneiformis (fc), the crus cerebri (CC) and pyrimid (P) appear notably dark. In contrast nuclear areas with large or compacted neuronal cell bodies appear rather transparent under transillumination. Examples of this situation are: interstitial nucleus of the stria terminalis (nist), suprachiasmatic (nsc), supraoptic (nso), paraventricular (npv) and habenular (hl) nuclei. Other areas containing both neuronal cell bodies and myelinated fibres display different degrees of opacity according to the relative density of these two elements. Examples of this are provided by the following areas: accumbens (a), caudate-putamen (cp), and globus pallidus (gp). In some situations the fibres are sparsely distributed in bundles all along the nuclei such as in the substantia nigra, pars reticulata (snr, Figure 21) or in the nucleus vestibularis spinalis (vsp) (Figure 30). In the corpus striatum the fibre bundles acquire a radial organization and its appearance justifies the classical nomenclature. The differential density of these fibres, denser in the globus pallidus allows the sharp distinction between this nucleus and the caudate putamen (gp, cp, Figures 12–14). Finally, a fibre tract might be accompanied by a nuclear organization such as in the case of the nucleus solitarius and tract (nts, TS, Figure 32) or the mesencephalic root of the trigeminal nerve (ntm, Figure 30). The reticular formation rarely forms a defined nuclear organization and the recognition of reticular nuclei is further complicated by the dense mesh of myelinated fibres present at those levels. Approximate limits for some of these nuclei are represented in Figures 25–34.

For those areas in which the researcher does not find clear-cut limits based purely on differential transparency of the slice, it is recommended that micrographs be produced of the same slice before and after staining with methylene blue, as described in Section 3.2 for comparison. This exercise will permit landmarks to be established on the unstained tissue slice and thus allow quicker microdissection.

Some levels of the rat CNS are easier than others. For instance most telencephalic, diencephalic, and mesencephalic portions do not offer major problems either for cutting or for nuclear recognition. In these planes special care has to be taken when the median eminence is being dissected. For this the pia mater covering the ventral aspect of the hypothalamus should be carefully removed. Even in these circumstances the median eminence tends to appear only attached to one side of the hypothalamus or with a different thickness from the rest of the section. As pointed out above, the medulla oblongata and particularly the pons, appears rather dense on fresh slices. In addition, peripheral structures here seem to suffer more than in the rest of the NCS, thus the cochlear nuclei are frequently lost unless the brain is removed from the skull with extra care and the cutting of the cerebellar peduncles is performed far away from the stem. However, a number of structures in the medulla oblongata such as the substantia gelatinosa of the nucleus spinalis of the trigeminal nerve, the nucleus of the hypoglossal nerve (Figure 33), the nucleus of the tractus solitarius (Figure 33), the dorsal nucleus of the vagus (Figure 34), the nucleus cuneafornis and gracilis (Figure 34), the olive complex (Figure 34) etc., are visualized with remarkable contrast.

4.2 Identification of microanatomical areas in fresh tissue slices of the rat brain

The diagrams accompanying the micrographs of fresh tissue slices represent the main microanatomical areas seen at the plane of section at low magnification. The approximate plane of section from the ventral aspect is represented in some figures by broken lines in inset micrographs of the entire rat brain and brain stem.

4.2.1 *Abbreviations to Figures 6–35*

a	nucleus accumbens
aa	area amygdaloidea anterior
ab	nucleus amygdaloideus basalis
abp	nucleus amygdaloideus basalis posterior
ac	nucleus amygdaloideus centralis
aco	nucleus amygdaloideus corticalis
aD	nucleus accessorius Darkschewitsch
AF	fasciculus anterioris
al	nucleus amygdaloideus lateralis
am	nucleus amygdaloideus medialis
amb	nucleus ambiguus
apo	nucleus amygdaloideus posterior
arc	fibrae arciformis
AS	aqueductus cerebri (Sylvii)
CA	commissura anterior
ca	nucleus interstitialis commissurae anterioris
CAI	capsula interna
CC	crus cerebri

CCA	corpus callosum
CCS	commissura colliculorum superiorum
CF	commissura fornicis
cgm	nucleus corporis geniculati medialis
ci	colliculus inferior
cl	claustrum
cod	nucleus cochlearis dorsalis
CP	commissura posterior
cp	nucleus caudatus putamen
CT	corpus trapezoideum
cu	nucleus cuneiformis
cul	nucleus cuneiformis lateralis
dcgl	nucleus dorsalis corporis geniculati lateralis
DH	cornus dorsalis
DP	decussatio pyramidalis
ep	nucleus entopeduncularis
epy	ependyma
F	fornix
fC	fasciculus cuneiformis
FG	fasciculus gracilis
FH	fimbria hippocampi
FLM	fasciculus longitudinalis medialis
FMT	fasciculus mamillothalamicus
FR	fasciculus retroflexus
gr	nucleus gracilis
gp	globus pallidus
HI	hippocampus
hl	nucleus habenulae lateralis
hm	nucleus habenulae medialis
io	nucleus olivaris inferior
ip	nucleus interpeduncularis
lc	locus coeruleus
LF	fasciculus lateralis
LGI	lamina granularis interna bulbi olfactorii
lha	area lateralis hypothalami
LL	lemniscus lateralis
lld	nucleus lemnisci lateralis dorsalis
llv	nucleus lemnisci lateralis ventralis
LM	lemniscus medialis
LMIO	lamina medullaris interna bulbi olfactorii
LMO	lamina molecularis bulbi olfactorii
LV	ventriculus lateralis
ME	median eminence

MFB	fasciculus medialis prosencephali (medial forebrain bundle)
mp	nucleus mamillaris posterior
na	nucleus arcuatus
ncs	nucleus centralis superior
ncu	nucleus cuneiformis
ndm	nucleus dorsomedialis
nha	nucleus hypothalamicus anterior
nhp	nucleus hypothalmicus posterior
nic	nucleus intercalatus
nist	nucleus interstitialis striae terminalis
nistd	nucleus interstitialis striae terminalis pars dorsalis
nistv	nucleus interstitialis striae terminalis pars ventralis
nlc	nucleus linearis caudalis
nlo	nucleus linearis oralis
npd	nucleus parabrachialis dorsalis
npmd	nucleus premamillaris dorsalis
npmv	nucleus premamillaris ventralis
npv	nucleus paraventricularis
npve	nucleus parabrachialis ventralis
nrd	nucleus reticularis medullae oblongatae pars dorsalis
nrp	nucleus reticularis paramedianus
nrv	nucleus reticularis medullae oblongatae pars ventralis
nsc	nucleus suprachiasmaticus
nso	nucleus supraopticus
ntd	nucleus tegmenti dorsalis Gudden
ntdl	nucleus tegmenti dorsalis lateralis
ntm	nucleus tractus mesencephali
nts	nucleus tractus solitarii
ntv	nucleus tegmenti ventralis Gudden
ntV	nucleus tractus spinalis nervi trigemini
nvm	nucleus ventromedialis
nIII	nucleus originis nervi oculomotorii
nV	nucleus originis nervi trigemini
nX	nucleus originis dorsalis vagi
nXII	nucleus originis nervi hypoglossi
oad	nucleus olfactorius anterior, pars dorsalis
oal	nucleus olfactorius anterior, pars lateralis
oam	nucleus olfactorius anterior, pars medialis
ol	nucleus tractus olfactorii lateralis
P	tractus corticospinalis
PCI	pedunculus cerebellaris inferior
PCM	pedunculus cerebellaris medius
PCMA	pedunculus corporis mamillaris

PCS	penduculus cerebellaris superior
pf	nucleus parafascicularis
ph	nucleus prepositus hypoglossi
pi	cortex piriformis
pol	nucleus preopticus lateralis
pom	nucleus preopticus medialis
pos	nucleus preopticus suprachiasmaticus
pt	nucleus paratenialis
pv	nucleus periventricularis thalami
r	nucleus ruber
rd	nucleus raphe dorsalis
re	nucleus reuniens
rgi	nucleus reticularis gigantocellularis
rh	nucleus rhomboideus
rl	nucleus reticularis lateralis
rm	nucleus raphe magnus
ro	nucleus raphe obscurus
rp	nucleus raphe pallidus
rpc	nucleus reticularis parvocellularis
rpo	nucleus raphe pontis
rpoo	nucleus reticularis pontis oralis
rtp	nucleus reticularis tegmenti pontis
S	nucleus septus
SAM	stratum album mediale colliculi superioris
SD	decussatio sensoralis
SGC	substantia grisea centralis
sgm	stratum griseum mediale colliculi superioris
sgs	stratum griseum superficiale colliculi superioris
sgV	substantia gelatinosa trigemini
SM	stria medullaris thalami
sn	substantia nigra
snc	substantia nigra zona compacta
snl	substantia nigra pars lateralis
snr	substantia nigra zona reticularis
SO	stratum opticum colliculi superioris
SR	sulcus rhinalis
ST	stria terminalis
sut	nucleus subthalamicus
tad	nucleus anterior dorsalis thalami
tam	nucleus anterior medialis thalami
tav	nucleus anterior ventralis thalami
TD	tractus diagonalis Broca
td	nucleus tractus diagonalis Broca

tl	nucleus lateralis thalami
tlp	nucleus lateralis posterior thalami
TM	tractus mesencephalicus nervi trigemini
tml	nucleus medialis thalami pars lateralis
tmm	nucleus medialis thalami pars medialis
TO	tractus opticus
TOL	tractus olfactorius lateralis
TOLD	tractus olfactorius lateralis, pars dorsalis
TOLI	tractus olfactorius lateralis, pars intermedia
TOLV	tractus olfactorius lateralis, pars ventralis
TOM	tractus olfactorius medialis
tpm	nucleus posteromedianus thalami
tpo	nucleus posterior thalami
tr	nucleus reticularis thalami
TS	tractus solitarius
TSV	tractus spinalis nervi trigemini
tu	tuberculum olfactorium
tv	nucleus ventralis thalami
tvd	nucleus ventralis thalami pars dorsalis
tvm	nucleus ventralis medialis thalami pars magnocellularis
vcgl	nucleus ventralis corporis geniculati lateralis
vm	nucleus vestibularis medialis
VH	cornus ventralis
VO	ventriculus olfactorius
vsp	nucleus vestibularis spinalis
zi	zona incerta
V III	ventriculus tertius
V IV	ventriculus quartus
V	nervus trigeminalis
XII	nervus hypoglossi

5. EXAMPLES OF MICRODISSECTED AREAS OF FRESH RAT BRAIN

Figure 36–39 illustrate the microdissection of some specific nuclear areas. Figure 36A represents the ventral aspect of the lower mesencephalon in a cryostat section stained with toluidine blue showing the subnuclear organization of the substantia nigra into pars compacta, reticulata and lateralis. In Figure 36B the pars compacta has been dissected from a fresh, unfixed, tissue slice at a similar level as in A, while the pars reticulata (r) and lateralis (l) remain behind. In Figure 36C only the pars reticulata has been removed. Note that in these preparations the entire corresponding subnuclei have been taken. Figure 37A–C represent different stages of the microdissection of a

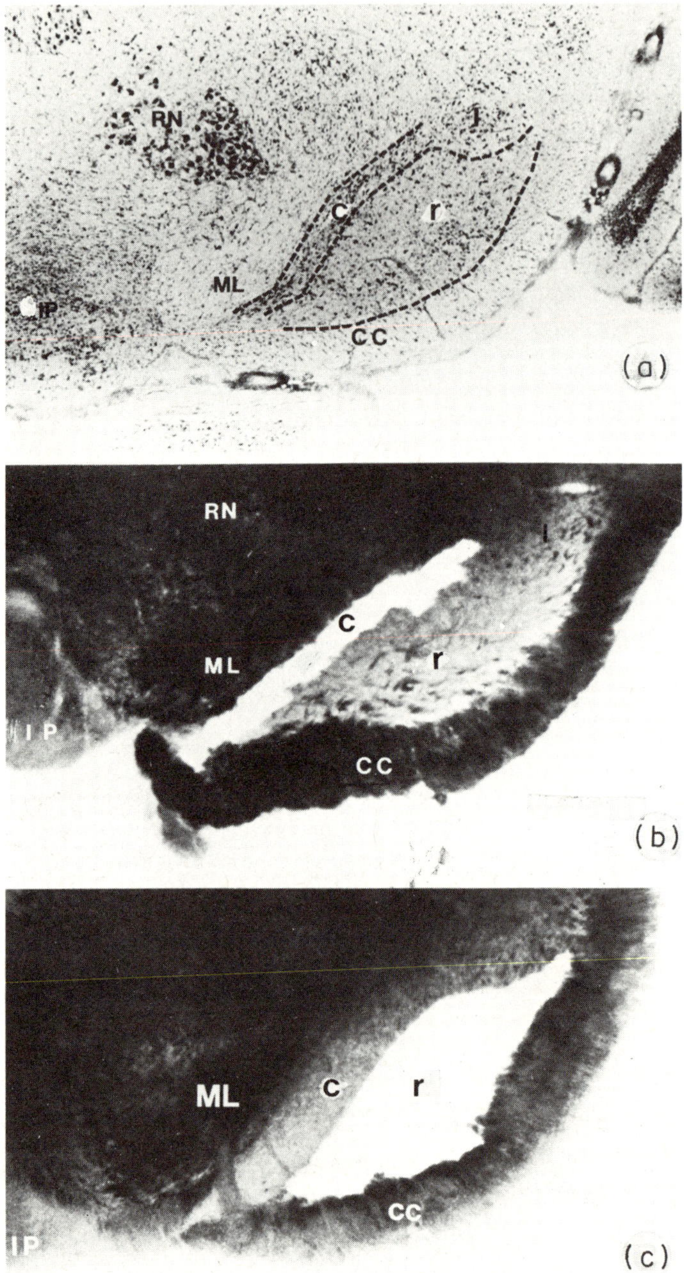

Figure 36 (a) Ventral aspect of the rat mesencephalon. 10 μm thick cryostat section, fixed with chloroform/methonoal and stained with toluidine blue.
(b) and (c) Approximately the same area as in (a). 250 μm thick fresh tissue sections. In (b) the substantia nigra pars compacta, c, has been dissected out and removed while in (c) the pars reticulata, r, of the substantia nigra from a similar preparation has been removed. RN, red nucleus; 1, substantia nigra pars lateralis; ML, medial lemniscus; CC, crus cerebri

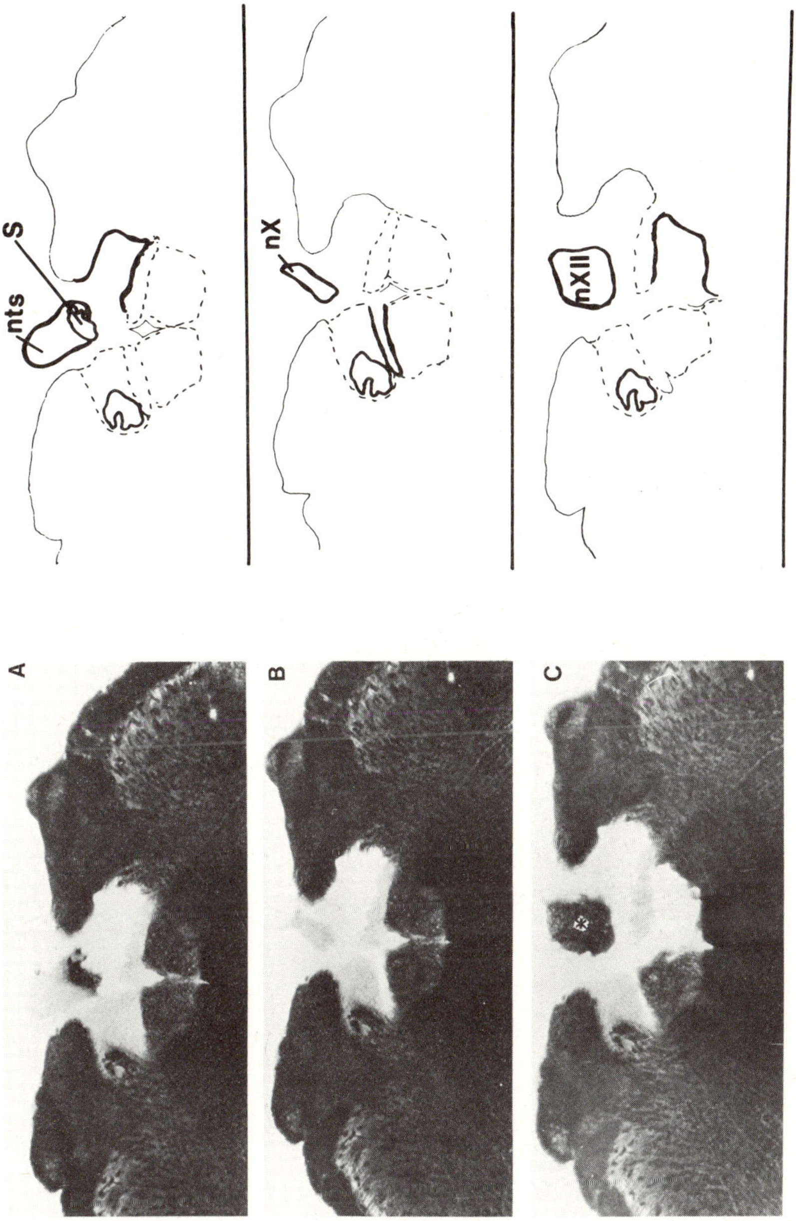

Figure 37

single tissue section of the lower medulla oblongata just anterior to the opening of the fourth ventricle. In Figure 37A the tractus solitarius (TS) and its nucleus (nts) has been removed on the right-hand side only. The dorsal nucleus of the vagus (nX) has been dissected out on the left-hand side in Figure 37B, while in Figure 37C the hypoglossal nucleus (asterisk, nXII) of

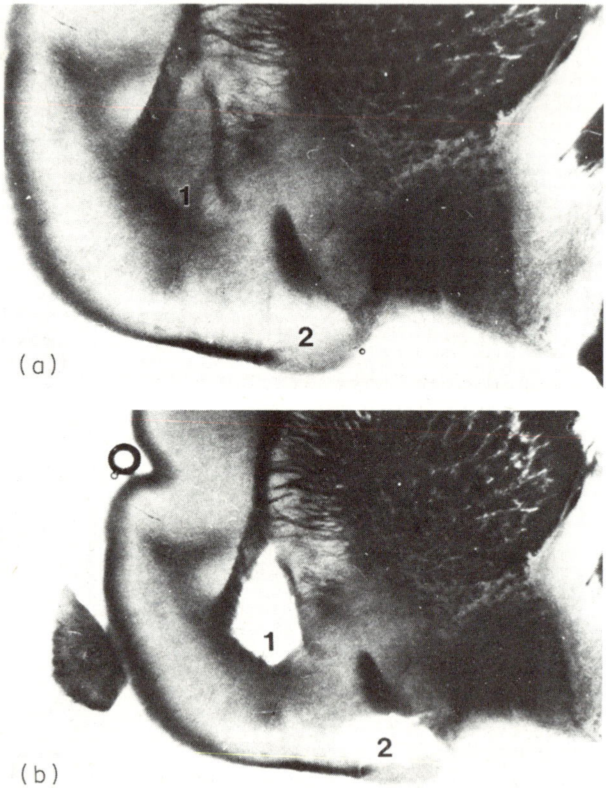

Figure 38

the right-hand side has been removed. Transparent dots in this nucleus and the one still remaining on the slice represent individual motoneurons.

The microdissected nucleus amygdaloideus lateralis (1) and olfactory lateralis (2) are observed in Figure 38B. The appearance of this nucleus in the intact tissue slice is seen in Figure 38A.

Figure 39A–C represent an example of how layered structures such as the olfactory bulb can be microdissected following precise histological landmarks denoting their synaptic organization. Fig 39A represents a Nissl stained section of the olfactory bulb (cresyl/violet) showing the various layers.

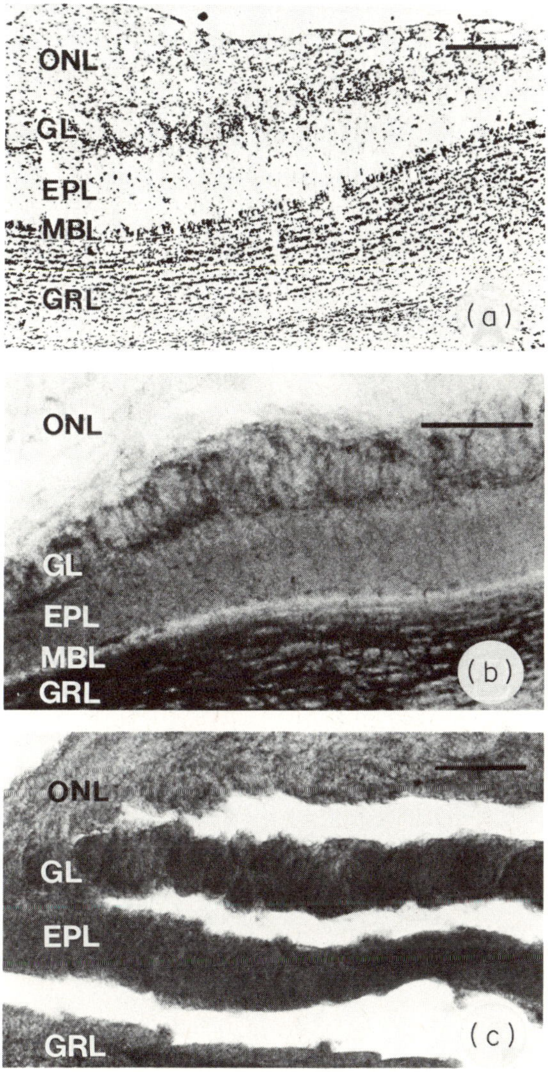

Figure 39

Olfactory bulb tissue slices intact are shown in Figure 39B and with the partially microdissected layers (c) (Jaffé and Cuello, 1980a).

6. APPLICATIONS OF THE TECHNIQUE

Microdissection of fresh tissue slices is a procedure which can be set up in most experimental laboratories. Equipment already available can be easily adapted for the purpose. Transilluminated slices of fresh brain display with remarkable clarity the microscopic organization of nuclear areas and fibre tracts. The use of fresh brain tissue slices can potentially be an economical and simple technique for the teaching of neuroanatomy. The tissue chopper is a relatively inexpensive piece of equipment which can provide a large number of sections of regular thickness in a short time. This is particularly convenient in neurochemistry when numerous samples need to be taken in a single session. As during sectioning with the tissue chopper the brain is exposed to the environment this operation should be carried out in a cold room. Alternatively the instrument should be placed in a refrigerated chamber thus permitting all the manipulations to be carried out at low temperatures. If this is not possible fresh sections can be obtained with a Vibratome as discussed in Section 2.2.2. Although the number and size of slices obtained is smaller, this instrument has the advantage that it can be placed next to the area where the microdissections and other experimental procedures are carried out. In addition thinner and better preserved sections are obtained. This instrument is the best option for fixed or fresh material which is destined for electron microscopy (for details see Priestley and Cuello, 1983).

Whether the final destiny of the microdissected material is the biochemical analysis of enzyme concentration, transmitter content or uptake of radiolabelled substances or their release, it is essential to confirm the microdissection sampling. For this, it is helpful to take as many micrographs as possible of fresh tissue slices before and after dissection. If the areas under study do not have very sharp landmarks under a transilluminated dissection microscope it is advisable to perform staining of the fresh tissue slices (see Section 3.2). The micrographs should also be correlated with micrographs of conventional Nissl stained sections taken at the same plane or, even better, with cryostat stained sections (approximately 30 μm thick) of the already dissected tissue slice. All these micrographs should be compared with available atlases (König and Klippel, 1963; Palkovits and Jacobowitz, 1974). There are difficulties in the interpretation of fresh tissue slices when compared with these atlases as stereotaxic planes are either presented schematically or as micrographs of fixed, stained sections. The aappearance of the slice is therefore different but not necessarily less informative. Another difficulty resides in the fact that planes of the stereotaxic atlases are taken following strict spatial orientation, a situation rather unpredictable when using the tissue chopper with very limited possibilities of orientation against the cutting arm. Nevertheless the above mentioned atlases represent the brain in sections perpendicular to the natural axis of the rat brain, which is the approximate result of simply placing the rat brain with the cortex downwards on the tissue chopper plate. The interested researcher when using this approach has to be alert to differences in the anterior–posterior planes of the section in dorsal *versus* ventral aspects of the same section. The difference should be similar in all the slices of the same series. In order to reconcile the classical view of the central nervous system from aniline or Weigert stained sections we have introduced a number of sequential anterior–posterior sections of the rat brain into this Chapter (see Section 4). We hope that they will be of some assistance to researchers beginning to perform microdissection of fresh tissue slices. Beginners should consult neuroanatomists when this expertise is readily available to validate their microdissections. One should perhaps also say that neuroanatomists lacking the understanding of the needs of neurochemists or neuropharmacologists sometimes find these procedures rather unorthodox. It is also true that a number of biochemists and pharmacologists who have been working on restricted areas of the brain are, on occasions, better prepared to deal with the microanatomy of a particular region than classical neuroanatomists.

What are the main advantages of the microdissection of fresh tissue slices? One of the advantages of this procedure is that the fine visualization of the microarchitecture of the brain using adequate transillumination allows a confident dissection following the clear-cut landmarks provided by dark fibre tracts and transparent grey matter organizations. This constitutes a real advantage as frozen slices are not transparent and orientation is based more on contour, ventricles, and major fibre tracts. Another advantage is that the procedure allows the removal of the entire given nuclear organization which makes the reproducibility of the sampling much better as nuclei vary greatly in the density of neuronal cell bodies and incoming nerve terminal throughout its limits. For instance the nucleus ventromedialis of the hypothalamus has the cell bodies concentrated in the inner core while most of the dendrites radiate outwardly with the bulk of the synaptic input being received in the superficial rim of the nucleus. Similarly, towards the inner parts of the nucleus accumbens a large number of indolamine and peptide containing fibre can be visualized by immunocytochemistry, but not on more lateral aspects. Therefore, a partial sampling of these and other nuclei will give a different answer according to the area taken on each occasion. In addition, the larger amount of material provided by the removal of the entire nuclei facilitates more accurate biochemical analysis. For example, the dopamine content of the rat substantia nigra pars compacta has been shown to be five-fold greater than that of the reticulata when microdissection is performed in fresh tissue slices (Cuello and Iversen, 1978), while only a two-fold difference is found between these two subnuclei by limited sampling of frozen sections. Ben-Ari *et al.* (1975) had a similar experience with the catecholamine content of microdissected nuclei of the amygdaloid complex when comparing both methodologies.

Another unquestionable advantage of the microdissection of fresh tissue section is that the membrane transport mechanisms are preserved while those in frozen tissue slices are lost. This is most important for the study of the incorporation of neurotransmitters in synaptically identified nuclear areas. Thus, the uptake of GABA and related amino acids could be explored in microdissected areas of the olfactory bulb containing predominantly glial cells, GABAergic nerve terminals or dendrites (Jaffé and Cuello, 1980b). Furthermore, fresh brain slices can be chemically stimulated by high K^+ and this property provided the intitial *in vitro* evidence for a dendritic release of dopamine from substantia nigra neurons (Geffen *et al.*, 1976; Cuello and Iversen, 1978). Microslices from fresh brain tissue obtained as described here, are sensitive to electrical field stimulation in superfusion systems (Aceves and Cuello, 1981). Microdissected samples from fresh tissue slices have been successfully applied to demonstrate the release of endogenous substances such as dopamine (Cuello and Iversen, 1978), serotonin (Reubi and Emson, 1978), substance P (Jessell and Iversen, 1977), and enkephalin (Iversen *et al.*, 1978).

Microdissection from fresh tissue slices has some important limitations. It is important to be familiar with the microanatomy of the area under study as orientation of the plane of sectioning cannot be as well controlled as in frozen sections. Another drawback is that some areas like the pons have very limited transparency with transillumination and although this can be partially corrected with thinner slices the detail of these sections is inferior to other parts of the central nervous system. In contrast with frozen cryostat sections, particularly with the tissue chopper although not with the Vibratome, some superficial structures can be torn during sectioning. This applies particularly to the brain stem.

Although for limited neurochemical exploration this procedure can be very rapid it has the disadvantage that the number of samples which can be taken from a single brain is regularly inferior to that from frozen sections where the stability of the tissue is obviously greater. It is for this reason that when undertaking microdissections from fresh brain slices realistic protocols should be set up, increasing the number of samples only when experience and facilities allow.

Finally, the researcher should judiciously choose a method compatible with experimental need when sampling microanatomical areas of the central nervous system. The present guidelines are intended as a framework for those wishing to use microdissection as a tool for answering specific questions in the neurosciences.

ACKNOWLEDGEMENTS

The authors would like to specially thank Miss Sara Patel for her skilful technical assistance. Thanks are also due to Mr Derek Auger for the preparation of micrographs and Miss Paula Cuello for illustrations. The contribution from Dr Erica Jaffé is also appreciated along with the dedicated and efficient secretarial assistance of Mrs Ella Iles.

REFERENCES

Aceves, J., and Cuello, A. C. (1981) 'Dopamine release induced by electrical stimulation of microdissected caudate-putamen and substantia nigra of the rat brain'. *Neuroscience*, **6,** 2069–2075.

Ben-Ari, Y., Zigmond, R. E., and Moore, K. E. (1975) 'Regional distribution of tyrosine hydroxylase, norepinephrine, and dopamine within the amygdaloid complex of the rat'. *Brain Res.*, **87,** 96–101.

Cuello, A. C., and Iversen, L. L. (1978) 'Interactions of dopamine with other neurotransmitters in the rat substantia nigra: a possible functional role of dendritic dopamine' in *International Symposium on Interactions Among Putative Neurotransmitters in the Brain.* (Eds. S. Garatinni, J. F. Pujol, and R. Samanin). Raven Press, New York, pp. 127–149.

Geffen, L. B., Jessel, T., Cuello, A. C., and Iversen, L. L. (1976) 'Release of dopamine from dendrites in rat substantia nigra.' *Nature*, **260,** 258–260.

Iversen, L. L., Iversen, S. D., Bloom, F. E., Vargo, T., and Guillemin, R. (1978) 'Release of enkephalin from rat globus pallidus *in vitro*.' *Nature,* **271,** 679–681.

Jaffé, E. H., and Cuello, A. C. (1980a) 'The distribution of catecholamines, glutamate decarboxylase and choline acetyltransferase in layers of the rat olfactory bulb.' *Brain Res.*, **186,** 232–237.

Jaffé, E. H., and Cuello, A. C. (1980b) 'Release of γ-aminobutyrate from the external plexiform layer of the rat olfactory bulb: possible dendritic involvement'. *Neuroscience*, **5,** 1869–1869.

Jaffé, E. H., and Cuello, A. C. (1981) 'Neuronal and glial release of 3H-GABA from the rat olfactory bulb.' *J. Neurochem.*, **37,** 1457–1466.

Jessell, T. M., and Iversen, L. L. (1977) 'Opiate analgesics inhibit substance P release from rat trigeminal nucleus.' *Nature*, **268,** 549–551.

König, J. F., and Klippel, R. A. (1963) *The rat brain. A stereotaxic atlas.* Robert Krieger, New York.

Palkovits, M., and Jacobowitz, D. M. (1974) 'Topographic atlas of catecholamine and acetylcholineresterase-containing neurons in the rat brain. II. Hindbrain (mesencephalon, rhombencephalon).' *J. Comp. Neurol.*, **157,** 29–42.

Priestley, J. V., and Cuello, A. C. (1982) 'Electron microscopic immunocytochemistry: CNS transmitters and transmitter markers' in *Immunohistochemistry*. (Ed. A. C. Cuello). John Wiley, Chichester. (in press)

Reubi, J. C., and Emson, P. C. (1978) 'Release and distribution of endogenous 5HT in rat substantia nigra.' *Brain Res.*, **139,** 164–168.

Zigmond, R. E., and Ben-Ari, Y. J. (1976) 'A simple method for the serial sectioning of fresh brain slices and removal of identifiable nuclei from stained sections for biochemical analysis.' *J. Neurochem.*, **26,** 1285–1287.

Brain Microdissection Techniques
Edited by A. C. Cuello

CHAPTER 3

Grid Microdissection of Human Brain Areas

ICHIRO KANAZAWA

Department of Neurology, Institute of Clinical Medicine, University of Tsukuba, Ibaraki-ken 305 Japan

CONTENTS

1. Introduction	127
2. Storage of frozen human brains	129
3. Preparation of sections of human brain areas	134
4. Procedures of microdissection	137
4.1 'Grid' microdissection	139
4.2 'At random' microdissection	150
5. Determination of volume of tissue samples	150

1. INTRODUCTION

In the nervous system the distribution of neurotransmitters or their synthesizing enzymes is one of the most attractive studies in neuroscience. Generally speaking, it is reasonable to expect that transmitter candidates and their enzymatic markers will be highly concentrated in nerve terminals. Indeed recent electronmicroscopical immunohistochemical techniques have revealed, for example, that glutamic acid decarboxylase (GAD), the γ-aminobutyric acid (GABA) synthesizing enzyme, is localized in certain populations of nerve terminals in the cerebellum (McLaughlin *et al.*, 1974), in the dorsal horn of the spinal cord (McLaughlin *et al.*, 1975) and in the substantia nigra (Ribak *et al.*, 1976). Moreover, choline acetyltransferase (acetylcholine synthesizing enzyme, Hattori *et al.*, 1977), tyrosine hydroxylase (catecholamine synthesizing enzyme, Pickel *et al.*, 1975), and neuropeptides such as substance P (Pickel *et al.*, 1979; Cuello *et al.*, 1979) have also been shown to be highly localized in nerve terminals in the central nervous system. On the other hand, it is fairly difficult to detect transmitter markers in nerve cell soma even with immunohistochemical methods. Since the central nervous system is composed of well-organized networks of fibre bundles connecting different nuclei, it is reasonable to assume that substances which are highly concentrated in certain nuclei and which show an uneven distribution pattern are present in nerve terminals. Clearly this assumption requires further support from pharmacological and physiological evidence, yet an elucidation of the distribution at the histological level of any compound in the central nervous system is still an essential step in a study of a transmitter candidate. For example, the finding that dopamine was highly concentrated in the striatum (Sano *et al.*, 1959) led later to the demonstration of the existence of the dopaminergic nigro-striatal fibre pathway (Anden *et al.*, 1964).

Studies on the distribution of transmitter markers in the nervous system can be categorized as follows:

1. Gross regional distribution studies. In this type of study anatomically identified nuclei such as cerebral cortex, striatum, substantia nigra etc., are grossly dissected as samples for biochemical analysis. Although these biochemical studies provide important information, they are not entirely satisfactory from the histological point of view.
2. Histochemical distribution studies. Enzymatic histochemistry (for example, acetylcholinesterase), fluorescence histochemistry (for example, catecholamines), or immunohistochemistry (for example, neuropeptides or enzymes) are well-documented methods used in neuroscience for microanatomical localization studies. Especially remarkable are recent advances in immunohistochemistry for the elucidation of the distribution of neuropeptides and marker enzymes in the nervous system (Hökfelt *et al.*, 1978). Moreover, some of these methods allow the localization of transmitter markers at the electron microscopical level (see above). However, these morphological methods are not entirely satisfactory from the biochemical (quantitative) point of view.
3. Fine regional distribution studies. This is a method which is able to reveal almost at the cellular resolution, the quantitative distribution of chemical substances. For this purpose, of course, it is necessary to use extremely sensitive analytical methods for the determination of substance(s) of interest. Lowry and his colleagues (Lowry and Passonneau, 1972) have developed an elegant enzymatic cycling method for the determination of concentrations of amino acids or other substances and Otsuka and Miyata (Miyata and Otsuka, 1972, 1975; Otsuka and Miyata, 1972) have applied this sensitive method for the quantitative mapping of GABA within the cat spinal cord. We have studied the distribution of GABA in the human basal ganglia using Otsuka's dissection method (Kanazawa *et al.*, 1973; Kanazawa and Toyokura, 1974, 1975). In this type of study tissue sections are microdissected into small pieces. This procedure is somewhat troublesome and time-consuming. However, it is quite useful for studying the distribution of a number of substances, including free amino acids, for which antisera have not been obtained so far.

Recently much attention has been paid to the biochemical changes in pathological human brains. For example, in Parkinson's disease dopamine and it synthesizing enzymes have been reported to be significantly decreased in the striatum. Moreover, in Huntington's chorea, GABA and its synthesizing enzyme have been shown to be depleted greatly in the substantia nigra (Perry *et al.*, 1973; Bird, 1976). In this Chapter, therefore, a microdissection method for the elucidation of the fine distribution of such substances in any area of the human brain will be described.

2. STORAGE OF FROZEN HUMAN BRAINS

As indicated in section 1 the most important advantage of the microdissection study is that biochemical results can be combined with morphological landmarks. It is therefore essential to keep the biochemical and morphological conditions of the human brain intact as far as possible.

From the biochemical point of view, the content of substances such as amino aicds, catecholamines or enzyme activities are influenced by various ante- and post-mortem factors. As an example of ante-mortem factors, GAD activity has been shown to be significantly decreased in patients dying with protracted illness (Spokes, 1979) and GAD and tyrosine hydroxylase activities and dopamine content have been reported to be decreased in the post-mortem period (Spokes and Koch, 1978; Puymirat *et al.*, 1979). These undesirable ante-mortem factors cannot be controlled in practical medicine. The only way to minimize such post-mortem factors is to put the cadaver into a refrigerator (4 °C) shortly after death until permission for autopsy is obtained. When the autopsy is started, the brain should be removed first and frozen as quickly as possible with the procedures described below.

From the morphological point of view, the most important point in handling the human brain is how to freeze it. For example, if a piece of brain is placed in a freezer for slow freezing, large numbers of ice crystals appear in the brain tissue, making the identification of the nucleus almost impossible when viewed under the microscope. Moreover, large pieces of frozen brain are difficult to handle when making thick slices suitable for cryostat sectioning. In order to avoid such disadvantages, we devised the following procedures for the treatment of post-mortem human brain.

1. In the autopsy room the whole fresh brain is cut at the midline into two halves using a long brain knife (dissecting blades, Brain No. 325, Feather Co.) (Figure 1).
2. Half the brain is brought to the cold room (4 °C) and the brain stem, with the cerebellum, is detached from the forebrain at the level of the rostral part of the midbrain by cutting with scissors or a knife (Figure 2).
3. Using a brain knife the forebrain is cut into two transverse sections at the level of the mammillary body (Figure 3).

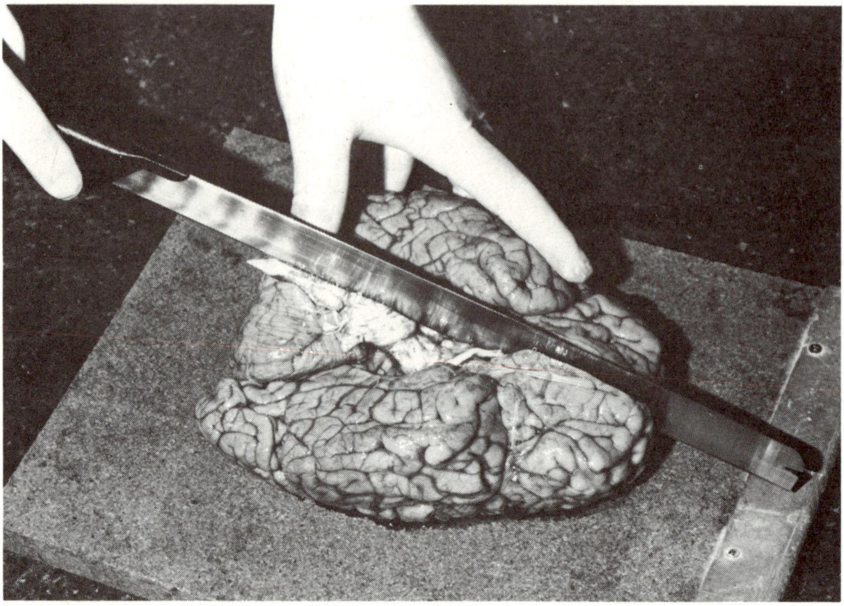

Figure 1 Cutting the human brain at the midline using a long brain knife. As shown in this photograph, the whole brain, including the brain stem and cerebellum is cut at the midline

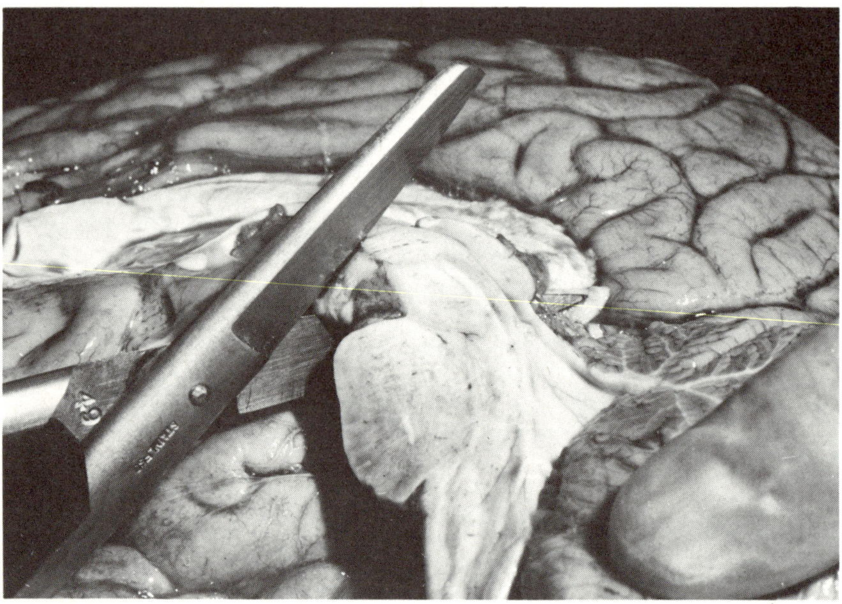

Figure 2 Detaching the brain stem. The brain stem with cerebellum detached from the forebrain at the level of the midbrain. If it is intended that the entire substantia nigra be preserved, a cut should be made at the extreme rostral level of the midbrain. If the whole hypothalamus is wanted in a single block a cut should be made at the middle of the midbrain

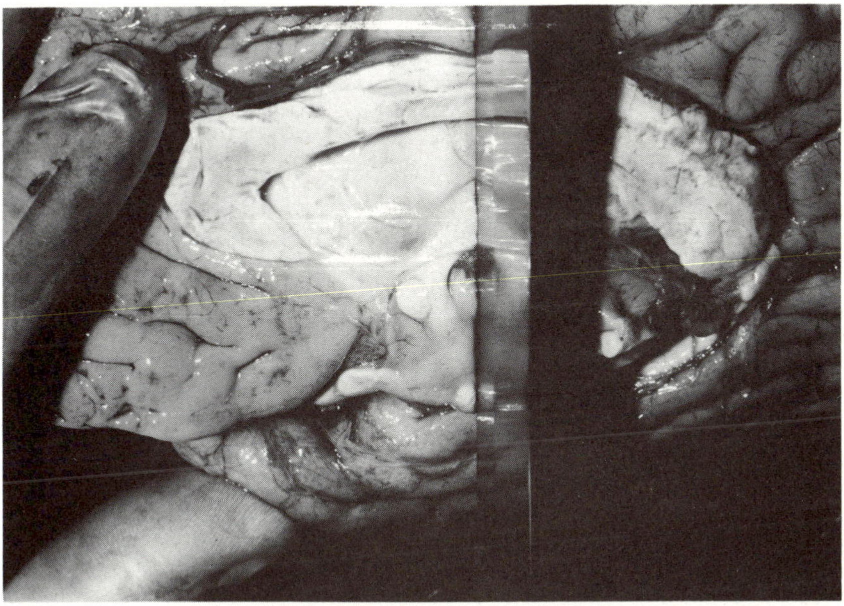

Figure 3 Cutting the forebrain. The forebrain is cut transversely into two at the level of the mammillary body

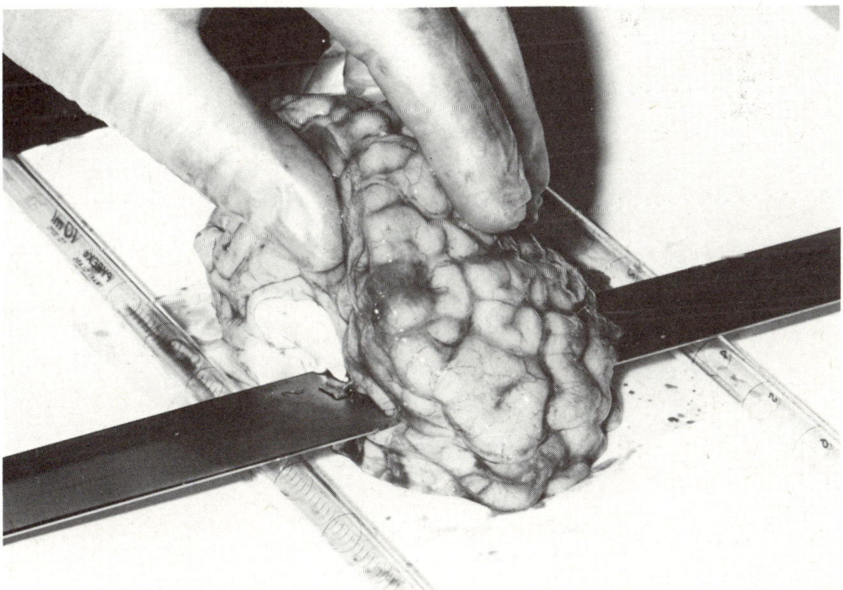

Figure 4 Making brain slices. The fresh brain is placed on a parafilm sheet and then sliced with a knife, using two glass pipettes as guides

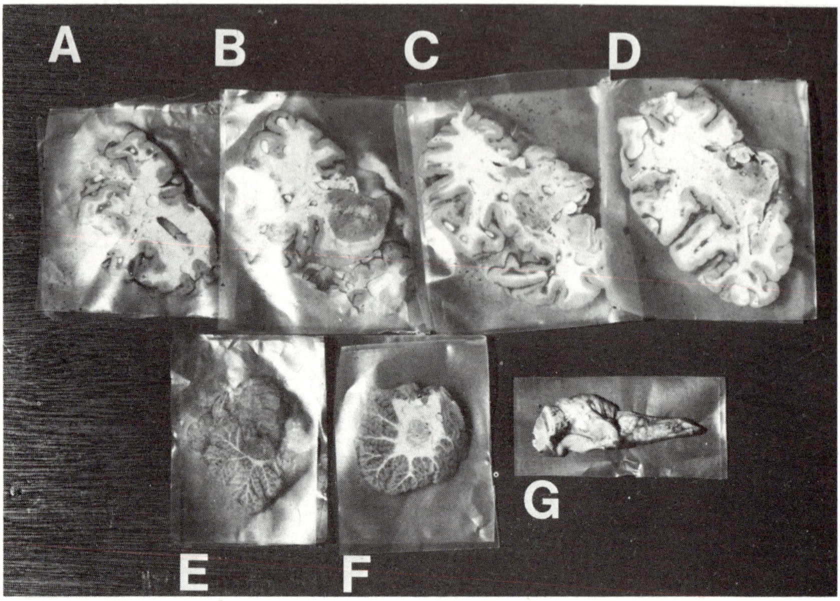

Figure 5 Fresh human brain slices (1 cm thick) are covered with parafilm on both sides. (A)–(D) forebrain slices from rostral to caudal. (E), (F) Sagittal slices of the cerebellum. (G) Brain stem

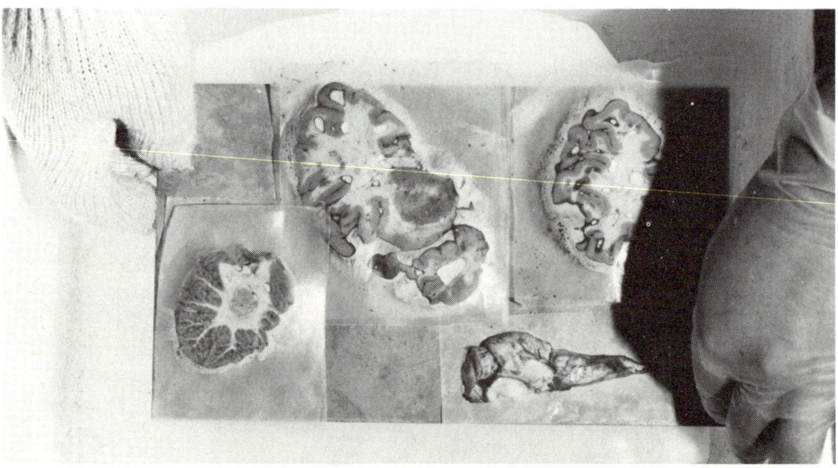

Figure 6 Freezing of brain slices. Slices are placed on a metal plate frozen in advance on dry-ice

4. With the cut surface down, the forebrain is placed on a parafilm sheet on a horizontal metal plate. The forebrain is held upright by hand during cutting to avoid deformations and two glass pipettes (8 mm or 10 mm in diameter) are placed parallel to the brain (see Figure 4). The brain is then cut with a dissecting blade which rests on these pipettes. Using this procedure 8 mm or 10 mm thick slices of fresh brain can be obtained quickly and easily. By repetition of this procedure several transverse sections of equal thickness can be obtained.

Figure 7 Plastic box for storage of brain slices. The diameter of the box is 15 cm. The lid of the box should be airtight

5. The cut surfaces of the brain slices are covered with parafilm sheets (i.e. both sides) (Figure 5), and then put on a dry ice-cold brass plate (Figure 6). Parafilm sheets are essential to prevent the tissue surface drying up or sticking to the dry ice-cold plate. It takes about five minutes for a freezing front to penetrate 5 mm.
6. It is necessary to turn the slice over when half the full thickness of the slice has been frozen. This method makes it possible to freeze large and thick brain slices quickly without the cracking or deformity produced when liquid nitrogen is used. Moreover, such a quick freezing process prevents ice-crystal formation.
7. The cerebellum is freed from the brain stem by cutting the three cerebellar peduncles with scissors. The cerebellum is put on a parafilm sheet with the midline surface down and it is then cut into 8 mm or 10 mm slices and frozen as above. The brain stem is frozen in the same way with the midline surface down.
8. The brain slices and parts of the brain thus obtained are packed in a plastic box (Figure 7) and kept in a deep freeze until required for use. In order to

keep the enzyme activities at their original level, the temperature of the freezer should be −80 °C or lower.

3. PREPARATION OF SECTIONS OR HUMAN BRAIN AREAS

The next step is to prepare frozen sections of human brain (150 μm or greater) for microdissection. A cryostat microtome is the most popular piece of equipment for this purpose. We use a Damon/IEC model CTD in our laboratory. This machine is suitable for making rather thick frozen tissue

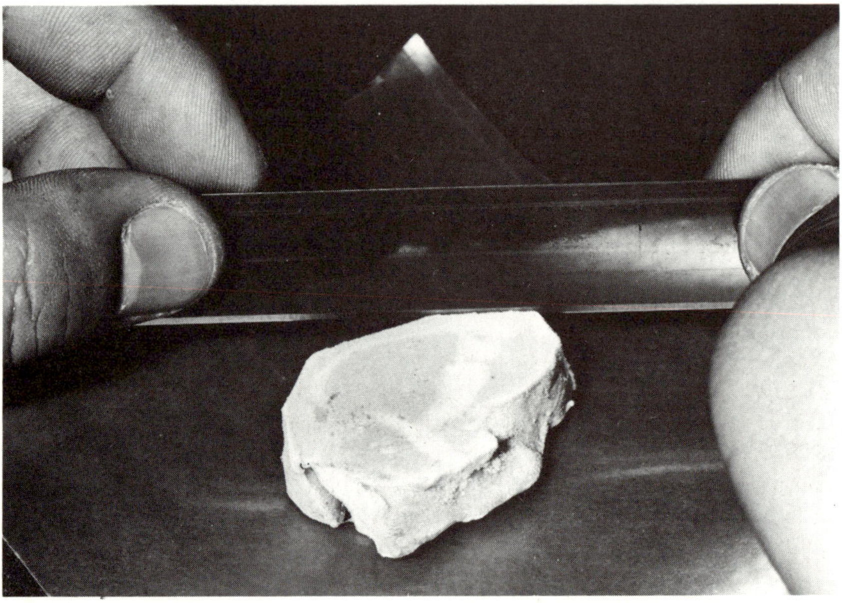

Figure 8 Thick brain slices are cut with a 'trimming blade' into tissue blocks of suitable size containing the area of interest (caudate nucleus and lenticular nucleus in this photograph)

sections because the blade of the microtome is set more horizontally than that of other types of machine where the cut sections immediately slip down. However, with this machine it is impossible to cut a tissue block larger than 45 mm × 35 mm because the sliding distance is limited. Therefore the first step has to be the trimming of the thick brain slices to a suitable size to include the area of interest. For this we use a Feather Co., Type S trimming blade for pathological operation (see Figure 8). During trimming, care should be taken to avoid the cracking of the tissue, and it is therefore advisable to keep the thick tissue slices in a cold room (4 °C) for at least ten minutes before trimming. The trimmed tissue block should then be mounted on a cryostat chuck with Tissue-Tek II (Lab-Tek Products). If the tissue block is small an ordinary cryostat chuck can be used. However, a tissue block in excess of 2.5 cm × 2.5 cm will require a specially made large chuck (Figure 9) otherwise the free edge of the tissue block will be easily deformed during the cutting procedure. A piece of exposed film should be mounted perpendicular to the cryostat chuck (Figure 9, 10) along with the tissue block. The brain tissue can then be sectioned at a thickness of 150 μm or 200 μm (Figure 11). This is achieved by setting the machine to cut 10 μm sections and turning the handle with 15 or 20 half-way turns to advance the tissue without cutting it. A

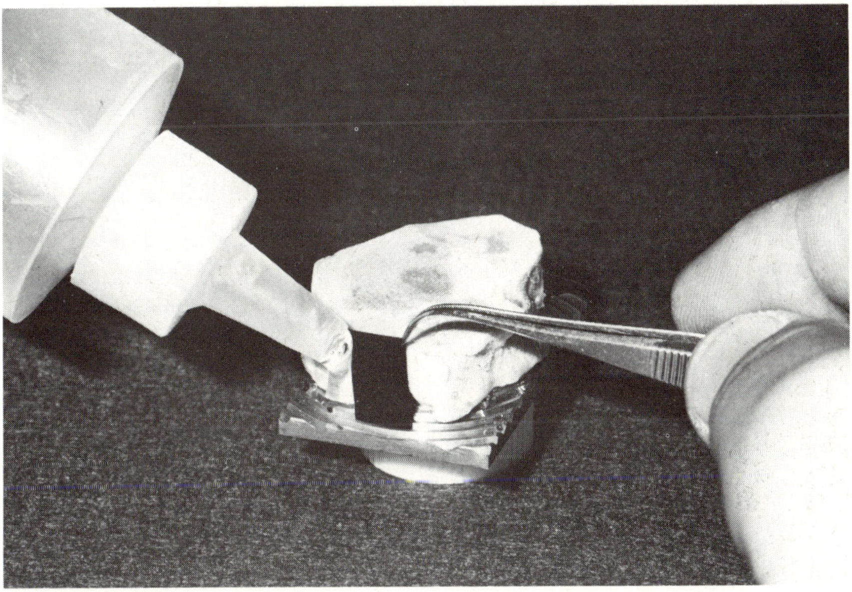

Figure 9 Mounting the tissue on a cryostat chuck. In this photograph a large tissue block is being mounted on a specially made large cryostat chuck with Tissue-Tek. A piece of exposed film is mounted perpendicularly

width of the strip of black exposed film will be cut along with the tissue and this can be used to calculate the exact thickness of the section (Figure 11 and (Section 5)). The tissue sections are transferred to a Perspex plate (Figure 12) with dry ice-cold forceps and then framed, if necessary, with strips of cover glass (see (iv) in Section 4.1 below). The size and thickness of the Perspex plate is arbitrary (we use plates of 40 mm × 50 mm and 1.5 mm thick). The frozen section on the Perspex plate is brought from the cryostat box to room temperature. The section is thus attached to the plate and can be photographed if necessary. It is convenient to use macrophotography apparatus (Olympus Co.) for this purpose (Figure 13).

Figure 10 Mounted tissue block and exposed film. The cryostat chuck with the tissue block and exposed film is put directly on to dry-ice for several minutes. Tissue-Tek, when frozen, is opaque and hard

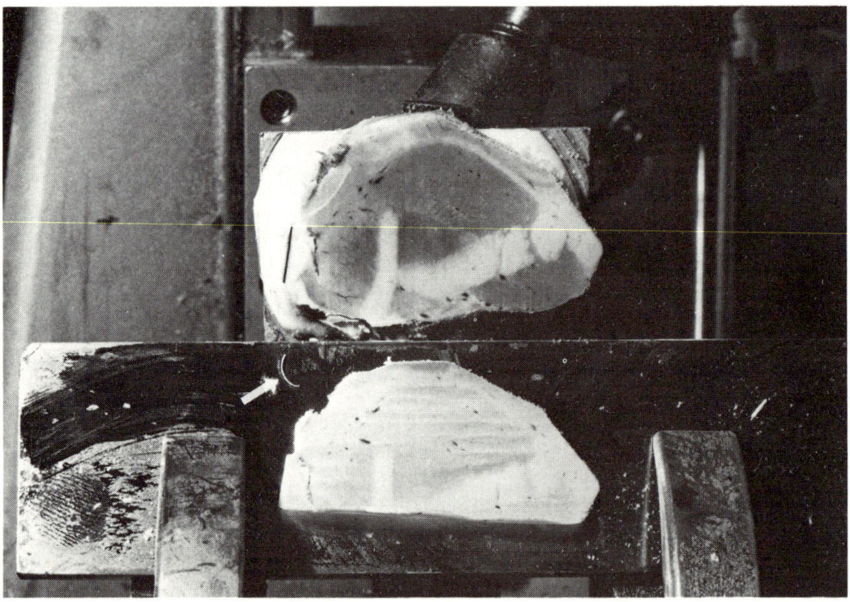

Figure 11 The tissue being sectioned at 200 μm thickness. The white arrow indicates the strip of exposed film being cut along with the tissue

If the thickness of the section is more than 200 μm it is difficult to dissect the tissue accurately with perpendicular, straight cuts. Conversely, if the thickness is less than 150 μm, it becomes difficult to remove the pieces of tissue following microdissection. Moreover, the thickness should be set according to the sensitivity of the analytical method being used.

The inner temperature of the cryostat microtome seems to be an important factor in preparing good frozen sections for microdissection. If the tissue block is large, the temperature should not be lower than −15 °C, otherwise

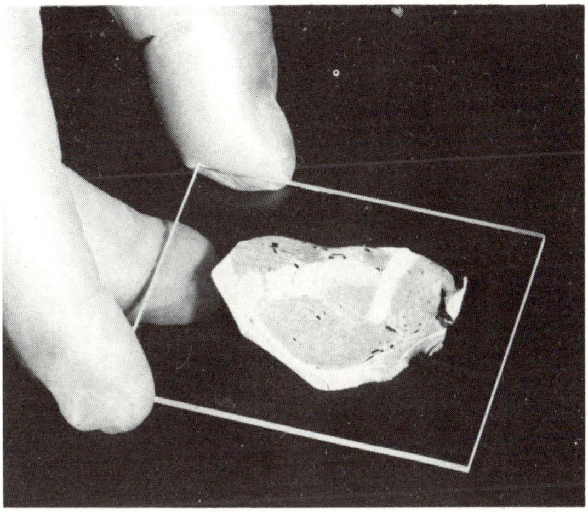

Figure 12 Frozen cryostat sections are placed on a Perspex plate using frozen forceps and then kept at room temperature for a couple of minutes. The tissue is thus attached to the plate

sections are easily wrinkled or striped (Figure 14). Another important factor in preparing good sections is the speed of cut. Slow but continuous moves of the microtome handle are required, especially when cutting large pieces of tissue.

For the precise microscopical identification of nuclei it is recommended that adjacent sections be cut at 50 μm thickness and placed on a glass slide which has been precoated with chrome–alum solution. For histological examination, the section is stained using, for example, the Klüver–Berrera staining method.

4. PROCEDURES OF MICRODISSECTION

We use two types of microdissection methods on these sections: 'grid' microdissection and 'at random' microdissection.

Figure 13 Macrophotography apparatus. The perspex plate is placed on the stage (indicated with arrow) and photographed at the desired magnification

Figure 14 Wrinkled cryostat section. This is an example of a poor section with obvious wrinkles

4.1 **'Grid' microdissection**

The 'grid' method was originally devised by Drs Miyata and Otsuka in 1972 when they were studying the distribution of GABA in the cat spinal cord (Miyata and Otsuka, 1972; Otsuka and Miyata, 1972). Since GABA is a small amino acid, it is difficult to prepare antibodies for immunohistochemical use. Therefore the 'grid' method is a very accurate way to elucidate the fine distribution of GABA in the nervous system. If the resolution is high enough it is possible to call the map obtained of GABA a 'quantitative histochemical map., Procedures for the grid microdissection are as follows:

1. Prepare a cold box. This specially ordered box was first described by Aprison and Hancock (1970) and modified by Miyata and Otsuka (1972) (Figure 15). The inner temperature is controlled by placing dry ice in the gutters surrounding the metal wall of the box (Figure 16). A blueprint of the box is given in Figure 17.
2. Set a binocular microscope. For microdissection it is necessary to use a binocular microscope with a long working distance (i.e. more than 15 cm) and various powers of magnification. For this purpose the model MTX Olympus binocular microscope with a long pole and arm is suitable (Figures 18 and 19).

Figure 15 Frontal view of the cold box used for microdissection of frozen sections

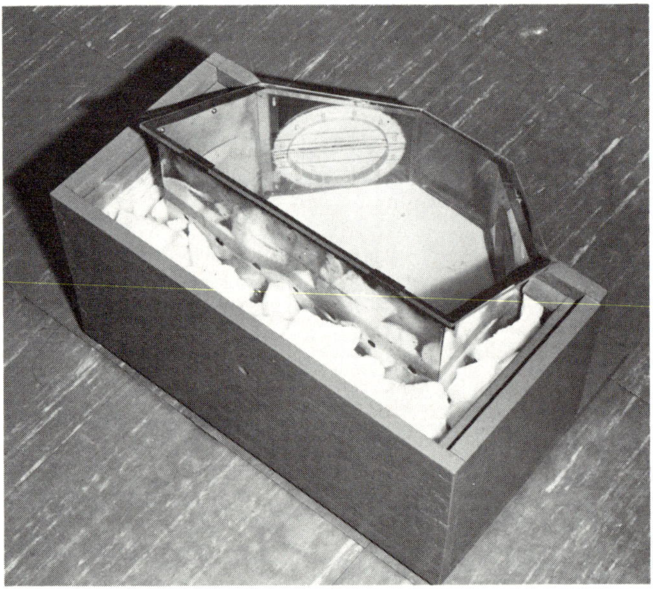

Figure 16 View of cold box from above. The gutters surrounding the metal walls of the box are filled with dry-ice

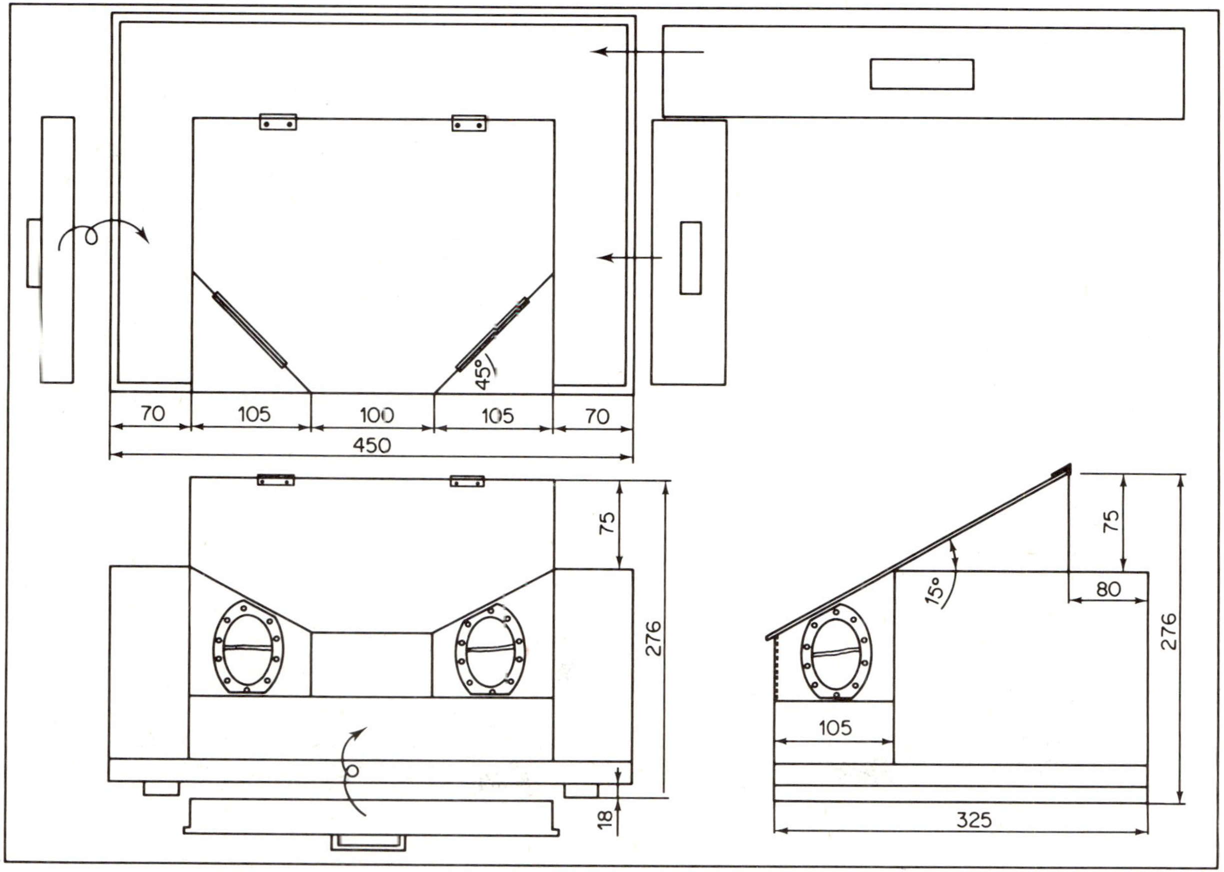

Figure 17 Blueprint of cold box. Dimensions are expressed in millimetres. The main lid and front panels of the box are made of Perspex. The inner walls and the bottom of the box are made of stainless steel. The outer box and gutter lids are wooden. Rubber cuffs are set into the round windows

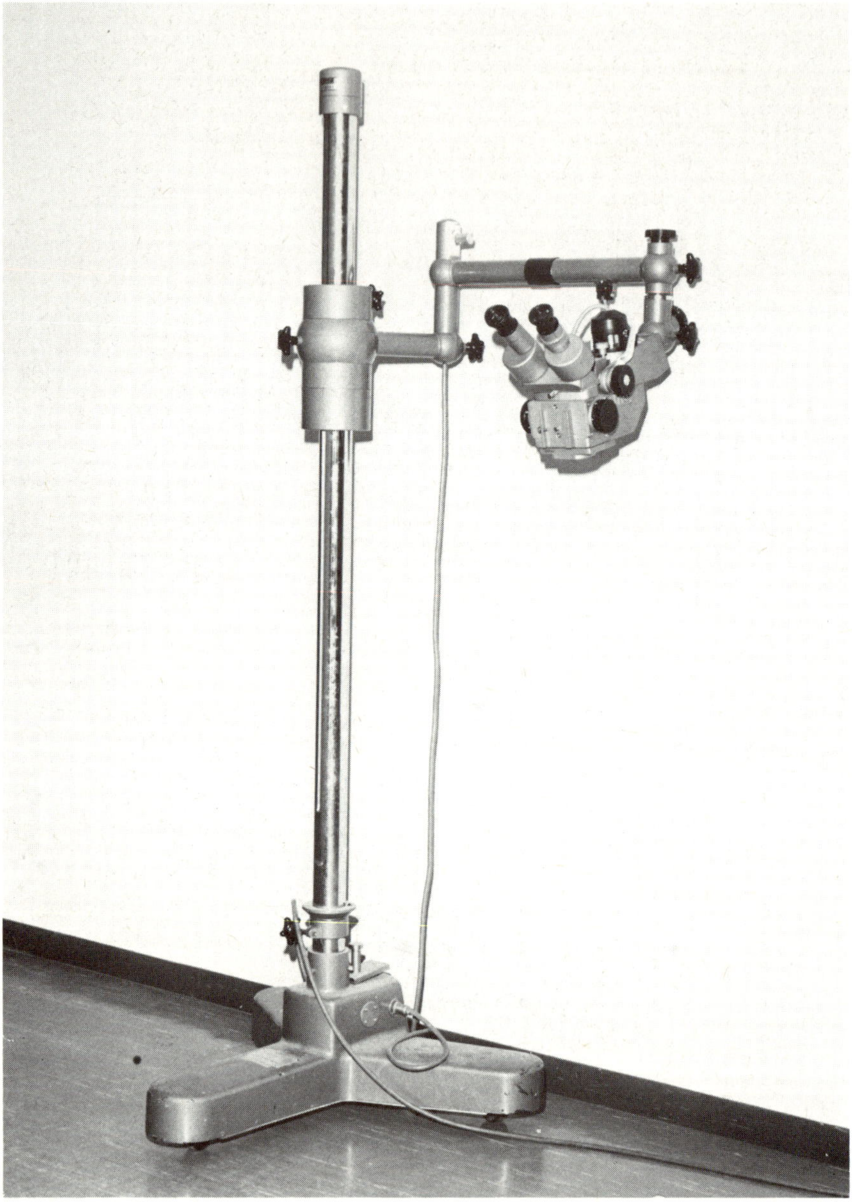

Figure 18 Olympus binocular microscope with long pole and arm as used for microdissection. (Model MTX)

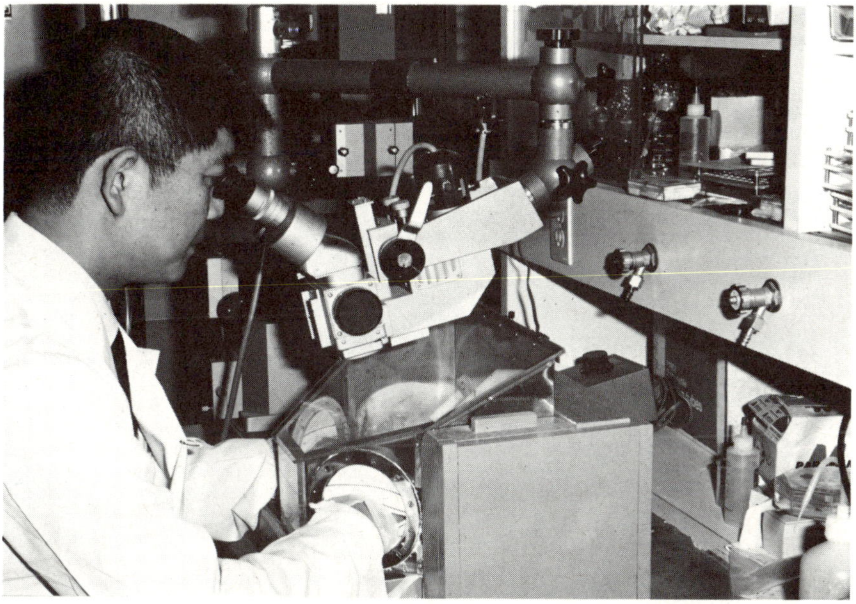

Figure 19 The microscope is set as shown in the photograph for viewing the tissue section through the Perspex cover of the cold box, at appropriate magnifications

Figure 20 A suitable blade can be made by breaking a frozen razor blade

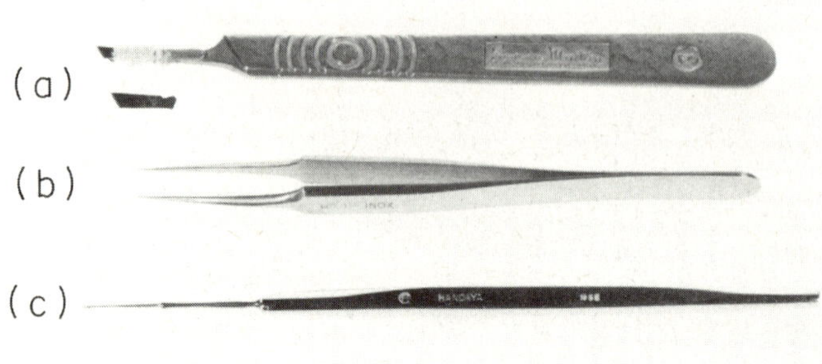

Figure 21 Small instruments for microdissection: (a) Broken razor blade tied to the tip of a blade holder. (b) Fine tipped forceps. (c) Bowman's needle

Figure 22 Microscale grid. Scale printed on a 4 cm × 5 cm glass slide at 500 μm intervals. For convenience the lines at every 5 mm intervals are thicker

3. It is advisable to wear rubber or plastic gloves when working in the cold box.
4. All small instruments should be placed in the cold box,.

   (i) A broken razor blade. A thin razor blade (Feather Co.) is placed directly on dry-ice for a while in order to make it brittle. The blade can then be broken into small pieces by bending it while holding it

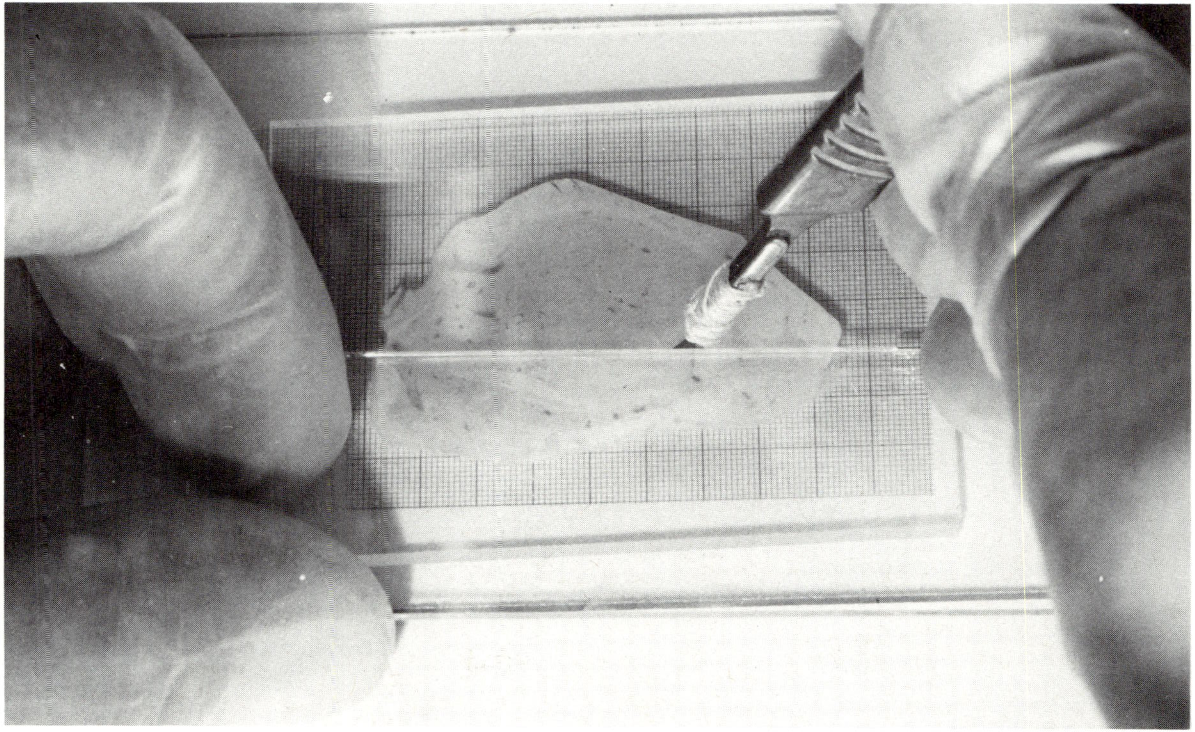

Figure 23 Cutting a straight line on a frozen section. The Perspex plate holding the tissue section is placed on the microscale and a cut is made using a glass slide as a guide

with pliers (Figure 20). The broken razor blade piece is then tied to the tip of a blade-holder (Figure 21A).

(ii) Fine tipped forceps (Figure 21B) used for picking up small samples.

(iii) Microscale (Figure 22). As a microdissection guide we use a specially made microscale which consists of a glass slide printed with crosswise lines at intervals of 500 μm.

(iv) Strips of glass cut from a cover slip. These glass strips are used to make the frame of the Perspex plate (see below).

(v) An ordinary glass slide. To be used as a guide when cutting the tissue sections in straight lines.

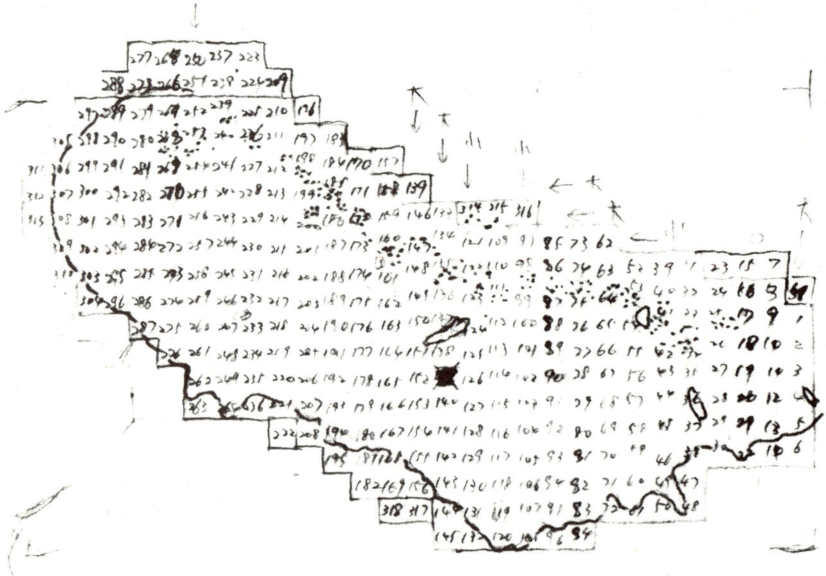

Figure 24 Rough sketch of cut section. Following complete microdissection of the tissue a rough sketch is made to allow easy identification of the position of each tissue section

5. It is useful to use a glass slide guide when using a microdissection knife. In order to cut straight tissue sections, the glass slide guide should be kept approximately 50 μm to 100 μm away from the tissue surface. The Perspex plate is framed, in advance, with strips of 0.3 mm thick and 5 mm wide glass cut from a cover slide. For framing, the glass cover strip is pressed lightly with the finger into the correct place on the Perspex plate. A drop of water is then pipetted into the narrow gap between the glass cover strip and the Perspex plate. Since the procedures are performed in

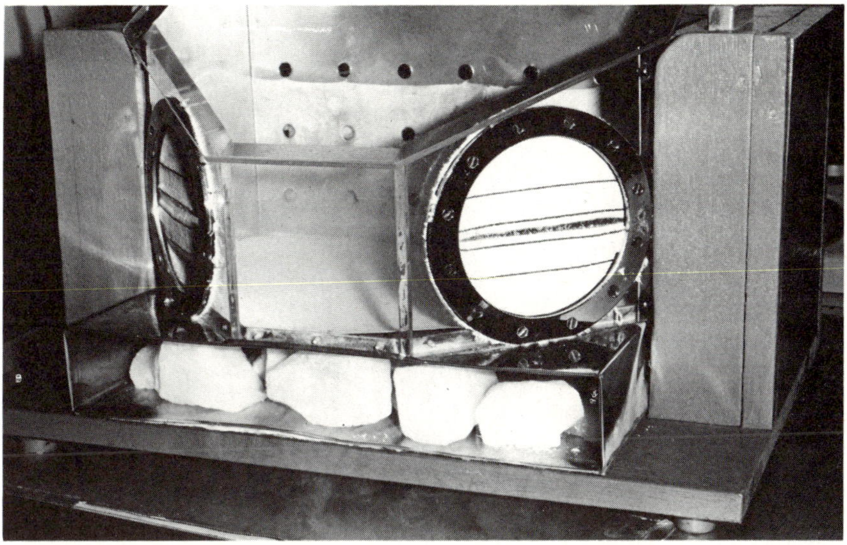

Figure 25 To lower the temperature inside the cold box to −50 °C dry-ice is placed in the compartment beneath the cutting plate

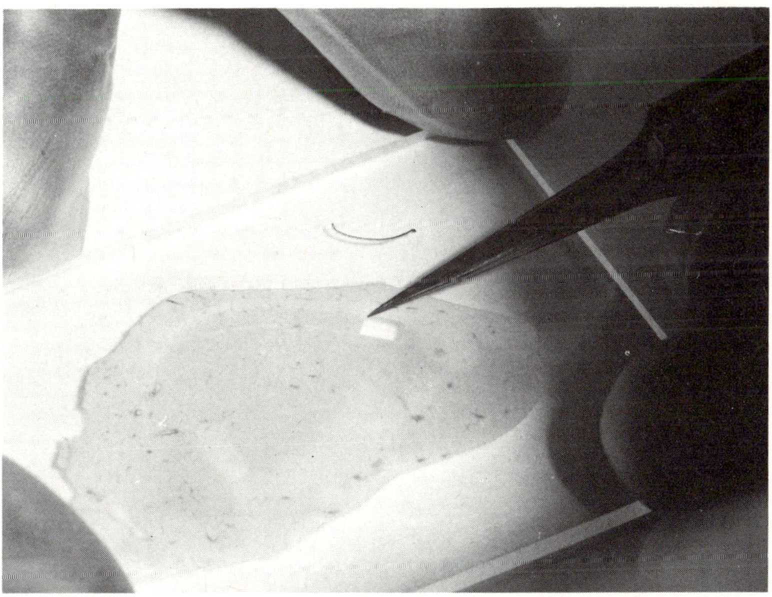

Figure 26 A small tissue block can be detached from the Perspex plate by gently pushing it with the tip of the forceps

the cold box (−20 °C) the water spreads by capillary action and is quickly frozen. In this way the strips of glass become bonded to the plate with ice. Alternatively it is possible to glue the glass strips to the plate using a chemical adhesive.

6. Place the microscale underneath the Perspex plate and attach them with ice as described above (Figure 23).
7. Under the binocular microscope and using the glass slide as a guide, the tissue section is initially cut in a straight line with the broken razor blade (Figure 23). During this procedure the first straight line should be drawn as exactly as possible following the line on the microscale.
8. By parallel shift of the glass slide successive straight cuts are performed at intervals of 500 μm or 1000 μm. The Perspex plate is then turned at right angles and a series of straight cuts are made in the same way. Thus the area of interest is microdissected into 500 μm or 200 μm squares. Take care not to move small pieces of tissue during cutting. To ensure successful cutting it is important to keep the temperature of the cold box between −20 °C and 25 °C.
9. The next step is to draw a rough sketch of the dissected area and to identify the position of each piece of tissue and number it. As an additional aid to identification it is advisable to record any characteristic marks (Figure 24).
10. By placing dry-ice in the lower compartment of the cold box, the inner temperature is reduced to around −50 °C (Figure 25).
11. By pushing the edge of the small blocks of tissue pieces (or 'flakes') with the fine tips of the forceps, the pieces are pulled off the Perspex plate (Figure 26). If the inner temperature of the cold box is not low enough, the tissue hardness will be insufficient to detach the pieces from the plate with this procedure. If the edge of the block of tissue is pushed with excessive force, many nearby pieces of tissue will be pulled off together, making identification of the location almost impossible.
12. When the small blocks of tissue are touched from above or from the side with the fine forceps, they readily stick to them and they can therefore be transferred easily to individual test tubes for biochemical determination (Figure 27). It is much easier to put the pieces into the test tubes if the test tubes have been frozen in the cold box in advance. Tissue blocks should be placed in the bottom of the tubes for ease of homogenization. Care should be taken during the transfer of the blocks to the test tubes otherwise some pieces may be lost. Static electricity sometimes pulls the blocks from the handling forceps and there is also a tendency for the pieces of tissue to stick to other objects such as finger tips or the outer surface of the test tube.
13. When all the small tissue blocks have been transferred to the test tubes, the Perspex plate is brought to room temperature and the marks cut on it

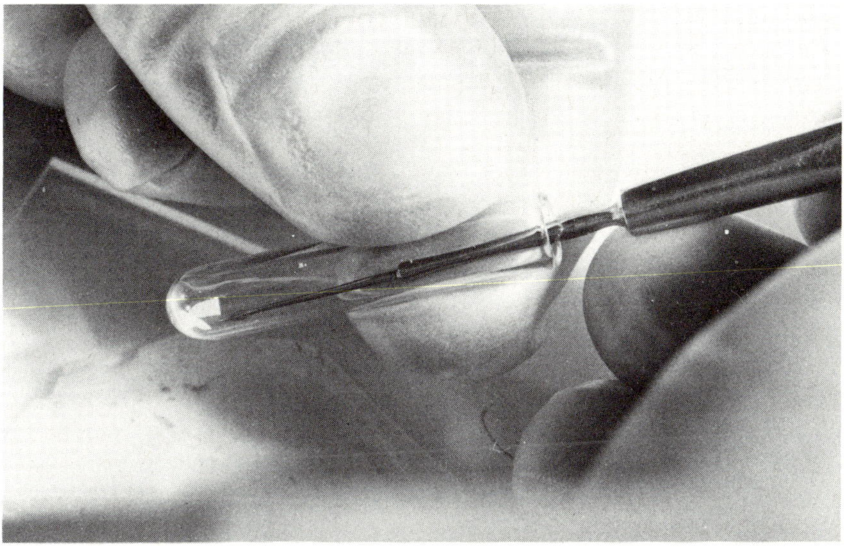

Figure 27 Transfer of tissue into small test tube. A tissue block can be transferred to the bottom of the individual test tube with the aid of a Bowman's needle or forceps. This procedure is carried out in the cold box

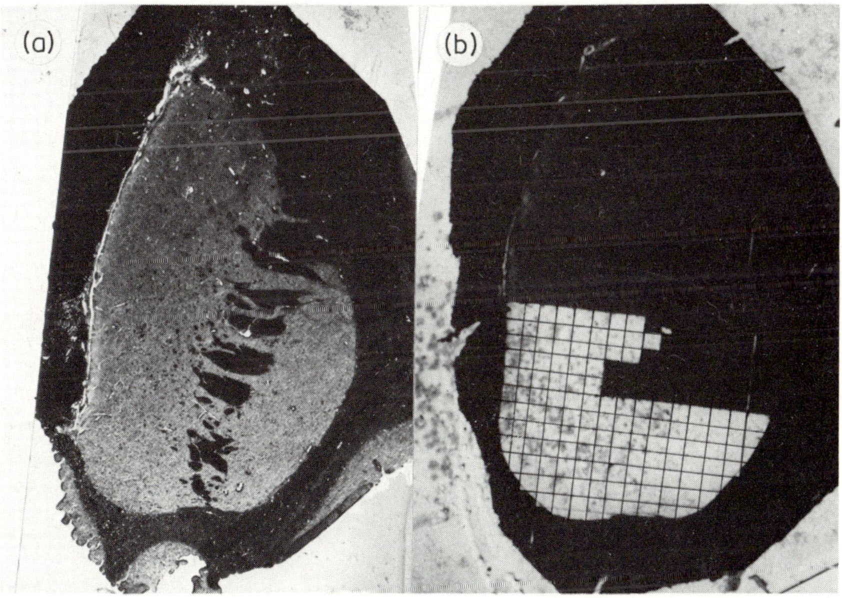

Figure 28 Enlarged photograph of a transverse section of human accumbens area: (a) Before microdissection. Unfrozen, unstained section. (b) After microdissection. Cut marks can be seen on the Perspex plate

are photographed with a macrophotography apparatus (Figure 28). This is essential for the calculation of the volume of the tissue pieces.

4.2 'At random' microdissection

This is a method in which the morphological landmarks are first cut with a single closed line and then the enclosed area is cut into small flakes of tissue at random. The advantage of this method is that it is easy to see the boundary of the nuclei under investigation when the map of the distribution of substance(s) is made. A typical example of this method is the study by Berger *et al.*, (1977) who microdissected the spinal cord of rabbit for the investigation of amino acid distribution. The size of each tissue flake depends on the sensitivity of the analytical chemical method and also on the total number of samples taken from one tissue section. It is usually difficult to handle more than 300 flakes in a day. We are presently doing microdissection of the human striatum using this method. After cutting the tissue section in the random fashion described above, a rough sketch is drawn on which each tissue flake is numbered according to the order in which it is picked up. Photographs should be taken before and after the microdissection. For the investigation of detailed histological landmarks or parameters (e.g. neuronal cell or glial cell number per limited square) we use an adjacent 50 μm thick section stained with a tradition staining method. After microdissection, each flake is detached from the Perspex plate by pushing the edge of the block with the sharp tip of a forceps or a blade of a Bowman's needle (Figure 22). During this procedure great care should be taken not to move adjacent tissue blocks. Each flake is picked up gently with fine-tipped forceps and put into individual, previously frozen, test tubes. The size of the test tube depends upon the homogenization and analytical methods employed. For most assays we use 5 mm × 25 mm Durham test tubes.

5. DETERMINATION OF VOLUME OF TISSUE SAMPLES

The samples taken are used for biochemical analyses. Since the concentration of substances or enzyme activities are usually expressed per protein content, an aliquot of each homogenate should be taken for protein determination. However, microdissected blocks of tissue or flakes are usually too small for the determination of both the substances of interest and the protein. In order to solve this problem Lowry and colleagues (Lowry *et al.*, 1956) devised a method of weighing extremely small tissue flakes which had been dissected from freeze-dried tissue sections using a fish pole calibrator. Unfortunately this method is too sophisticated for handling the large numbers of samples taken by the grid method.

Miyata and Otsuka (1972, 1975) developed a method to measure the volume of each block of frozen tissue from the photographs taken after microdissection. On the assumption that the tissue specific gravity is approximately 1.0, it is possible to calculate the wet weight of small samples from the volume. The first step is to measure the exact thickness of the tissue section. As described in Section 3, the perpendicularly placed strip of exposed film on the cryostat table is cut together with the tissue block. Under the microscope, the width of the thin film (Figure 26) is measured with a microscale. A calibrated scale in the eyepiece is used for the measurement of the width of the black film strip. This width represents the exact thickness of the tissue section. The second step is to measure the areas of the microdissected tissue block, using the enlarged photographs of the cut marks. It is important to take photographs of the microscale (Figure 22) on the same film as that used for the tissue section, using the same magnification enlargement. If the microdissection has been performed in random fashion, each circumscribed area on the photograph is cut with scissors and weighed. By dividing this weight with the standard weight obtained from photographs of the scale, the exact area of the piece of tissue can be calculated. By simple multiplication, the volume of the piece of tissue can be calculated. In the case of the grid method, since each block of tissue is rectangular in shape, each area can be directly determined by measurement with sliding calipers on the enlarged photographs.

REFERENCES

Ánden, N. E., Carlsson, A., Dahlstrom, Å., Fuxe, K., Hillarp, N. Å., and Larsson, K. (1964) 'Demonstration and mapping out of nigro-striatal dopamine neurons.' *Life Sci.*, **3,** 523–530.

Aprison, M. H., and Hancock, C. J. (1970) 'A controlled humidity constant temperature cold box for dissection of small frozen tissue sections.' *Exp. Physiol. Biochem.*, **3,** 39–48.

Berger, S. J., Carter, J. G., and Lowry, O. H. (1977) 'The distribution of glycine, GABA, glutamic acid and aspartate in rabbit spinal cord, cerebellum and hippothalamus.' *J. Neurochem.*, **28,** 149–158.

Bird, E. D. (1976) 'Biochemical studies on γ-aminobutyric acid metabolism in Huntington's chorea.' in *Biochemistry and Neurology*. (Eds. H. Bradford and C. D. Marsden) Academic Press, London, pp. 83–92.

Cuello, A. C., Jessell. T. M., Kanazawa, I., and Iversen, L. L. (1979) 'Substance P localization in synaptic vesicles in rat central nervous system.' *J. Neurochem.*, **29,** 747–751.

Hattori, T., McGeer, E. G., Singh, V. K., and McGeer, P. L. (1977) 'Cholinergic synapse of the interpeduncluar nucleus.' *Exp. Neurol.*, **55,** 666–679.

Hökfelt, T., Elde, R., Fuxe, K., Johansson, O., Ljungdahl, A., Goldstein, M., Luft, R., Efendic, S., Nilsson, G., Terenius, L., Ganten, D., Jeffcoate, S. L., Rehfeld, J., Said, S., Perez de la Mora, M., Possani, L., Tapia, R., Teran, L., and Palacios, R. (1978) 'Aminergic and peptidergic pathways in the nervous system with special reference to the hypothalamus' in *The Hypothalamus* (Eds. S. Reichlin, R. J. Baldessarini, and J. B. Martin) Raven Press, New York, pp. 69–135.

Kanazawa, I., Miyata, Y., Toyokura, Y., and Otsuka, M. (1973) 'The distribution of γ-aminobutyric acid (GABA) in the human substantia nigra.' *Brain Res.*, **51,** 363–365.

Kanazawa, I., and Toyokura, Y. (1974) 'Quantitative histochemistry of GABA in the human substantia nigra and the globus pallidus'. *Conf. Neurol. (Basel).*, **36,** 273–281.

Kanazawa, I. and Toyokura, Y. (1975) 'Topographical study of gamma-aminobutyric acid (GABA) in the human substantia nigra. A case study.' *Brain Res.*, **100,** 371–381.

Lowry, O. H., Roberts, N. R., and Lewis, C. (1956) 'The quantitative histochemistry of the retina.' *J. Biol. Chem.*, **220,** 879–892.

Lowry, O. H. and Passonneau, J. V. (1972) *A Flexible System of Enzymatic Analysis*, Academic Press, New York.

McLaughlin, B. J., Wood, J. G., Saito, K., Barber, R., Vaughn, J. E., Roberts, E., and Wu, J. Y. (1974) 'The fine structural localization of glutamate decarboxylase in synaptic terminals of rodent cerebellum.' *Brain Res.*, **76,** 377–391.

McLaughlin, B. J., Barber, R., Saito, K., Roberts, E., and Wu, J. Y. (1975) 'Immunocytochemical localisation of glutamate decarboxylase in rat spinal cord.' *J. Comp. Neurol.*, **164,** 305–321.

Miyata, Y., and Otsuka, M. (1972) 'Distribution of γ-aminobutyric acid in cat spinal cord and the alteration produced by local ischemia.' *J. Neurochem.*, **19,** 1833–1834.

Miyata, Y., and Otsuka, M. (1975) 'Quantitative histochemistry of γ-aminobutyric acid in cat spinal cord with special reference to presynaptic inhibition.' *J. Neurochem.*, **25,** 239–244.

Otsuka, M., and Miyata, Y. (1972) 'Application of enzymatic cycling to the measurement of gamma-aminobutyric acid in single neurons of the mammalian central nervous system,' in *Advances in Biochemical Psychopharmacology*, Raven Press, New York, Vol. 6, pp. 61–74.

Perry, T. L., Hansen, S., and Kloster, M. (1973) 'Huntington's chorea. Deficiency of γ-aminobutyric acid in brain.' *New Engl. J. Med.*, **288,** 337–342.

Pickel, V. M., Joh, T. H., and Reis, D. J. (1975) 'Ultrastructural localization of tyrosinehydroxylase in noradrenergic neurons of brain.' *Proc. Natl. Acad. Sci. USA.*, **72,** 659–663.

Pickel, V. M., Joh, T. H., Reis, D. J., Leeman, S. E., and Miller, R. J. (1979) 'Electron microscopic localisation of substance P and enkephalin in axon terminals related to dendrites of catecholaminergic neurons.' *Brain Res.*, **160,** 387–400.

Puymirat, J., Javoy-Agid, F., Gaspar, P., Ploska, A., Prochiantz, A. and Agid, Y. (1979) 'Post-mortem stability and storage in the cold of brain enzymes.' *J. Neurochem.*, **32,** 449–454.

Ribak, C. E., Vaughn, J. E., Saito, K., Barber, R., and Roberts, E. (1976) 'Immunocytochemical localization of glutamate decarboxylase in rat substantia nigra.' *Brain Rex.*, **116,** 287–298.

Sano, I., Gamo, T., Kakimoto, Y., Taniguchi, K., Takesada, M., and Nishinuma, K. (1959) 'Distribution of catechol compounds in human brain.' *Biochim. Biophys. Acta.*, **32,** 586–587.

Spokes, E. G. S., and Koch, D. J. (1978) 'Post-mortem stability of dopamine, glutamate decarboxylase, and choline acetyltransferase in the mouse brain under conditions simulating the handling of human autopsy material.' *J. Neurochem.*, **31,** 381–383.

Spokes, E. G. S. (1979) 'An analysis of factors influencing measurements of dopamine, noradrenaline, glutamate decarboxylase and choline acetylase in human post-mortem brain tissue.' *Brain*, **102,** 333–346.

Brain Microdissection Techniques
Edited by A. C. Cuello

CHAPTER 4

Large Section Cryomicrotomy in Human Neuroanatomy and Neurochemistry

STEN-MAGNUS AQUILONIUS, SVEN-ÅKE ECKERNÄS, and PER-GÖRAN GILLBERG

Department of Neurology, University Hospital, S-750 14 Uppsala, Sweden

CONTENTS

1. Introduction 155
2. Microtome sections of the human brain 156
   2.1 Heavy-duty microtomes 156
   2.2 Freezing and embedding of tissue 158
   2.3 Sectioning 160
3. Processing of sections 160
   3.1 Collecting samples for biochemical analysis 160
   3.2 Notes regarding safety precautions 163
4. Examples of application 166
   4.1 Biochemical mapping in human brain and spinal cord 166
   4.2 Regional biochemistry in whole human muscle 167
   4.3 Neuranatomy 167
   4.4 Neuropathology 167

1. INTRODUCTION

Whole body autoradiography of experimental animals was, in its preliminary state, introduced by Ullberg in 1954 and since then this technique has been of great importance in pharmacology and toxicology. Primarily, the development of elaborate equipment for cryosectioning was linked to the improved technique of autoradiography and heavy-duty cryomicrotomes (review by Ullberg, 1977).

More recently, however, large-section cryomicrotomy has found new areas of application. Thus, cryomicrotomy has become a valuable method in microdissectioning of tissue for neurochemical analysis (Aquilonius *et al.*, 1975, 1981) an illustrative technique in human topographic anatomy (Aquilonius and Eckernäs, 1980; Rauschning, 1980) and a valuable tool in pathology (Muhi *et al.*, 1982).

This chapter will present details of the technique of cryosectioning using a heavy-duty cryostat microtome. Furthermore, the procedure of punching out tissue samples from sections will be presented as well as some examples of the application of cryomicrotomy in neuroscience.

2. MICROTOME SECTIONS OF THE HUMAN BRAIN

2.1 Heavy-duty microtomes

In our laboratory we use a Swedish cryostat microtome (PMV 400, LKB 2250, LKB Sweden). This is without doubt the most widely used microtome for very large cryosections and to our knowledge only one other piece of similar equipment has been used in neuroscientific research (Kopp *et al.*, 1979). LKB 2250 is a large semi-automatic cryomicrotome (Figure 1) which

Figure 1 A LKB 2250 semi-automatic cryomicrotome

allows sectioning of 4500 mm × 1500 mm tissue blocks. After a slight modification the height of the specimen can be up to 2000 mm which permits sectioning of a whole human brain (Aquilonius and Eckernäs, 1980). The easily exchangeable stages are available in sizes down to 40 mm × 120 mm suitable for specimens such as human spinal cord. The range of sectioning can be varied from 1 μm to 999 μm. The microtome (Figure 2) standing in the cryostat works with a horizontally moving sledge, which is controlled automatically or by a knee control unit. The knife-holder is very stable (Figure 3). For nervous tissue sectioning the best results are obtained using a 20 ° edge-angle of the knife.

Figure 2 The microtome, normally standing within the cryostat

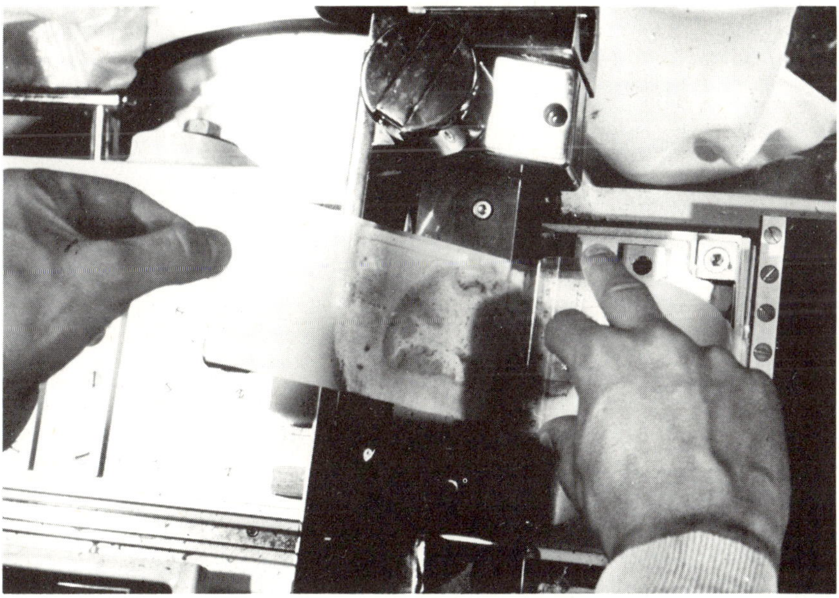

Figure 3 Section of a human brain hemisphere. The 200 μm section is taken up on tape

2.2 Freezing and embedding of tissue

For obvious reasons, the freezing technique must be adapted to the size of the specimen Small tissue volumes such as one segment of human spinal cord can be frozen in isopentane cooled with dry ice in acetone. In this way good histology is preserved and the slices can also be used for *in vitro* autoradiography. Large specimens such as a whole human brain must be frozen slowly in a

Figure 4 Human biceps muscle mounted in CMC and metal frame used when embedding the specimen

freezer (−20 °C) to avoid cracking, and with special precautions to avoid deformation (see below).

The frozen specimen is placed on a cooled (−70 °C) stage enclosed by a metal frame (Figure 4) and a semi-liquid gel of 4 °C carboxymethylcellulose (CMC) is added. The tissue–CMC block is frozen either in the freezer (large samples) or rapidly in hexane/dry ice to avoid thawing (small samples).

A special procedure for reproducible orientation of sections in a brain

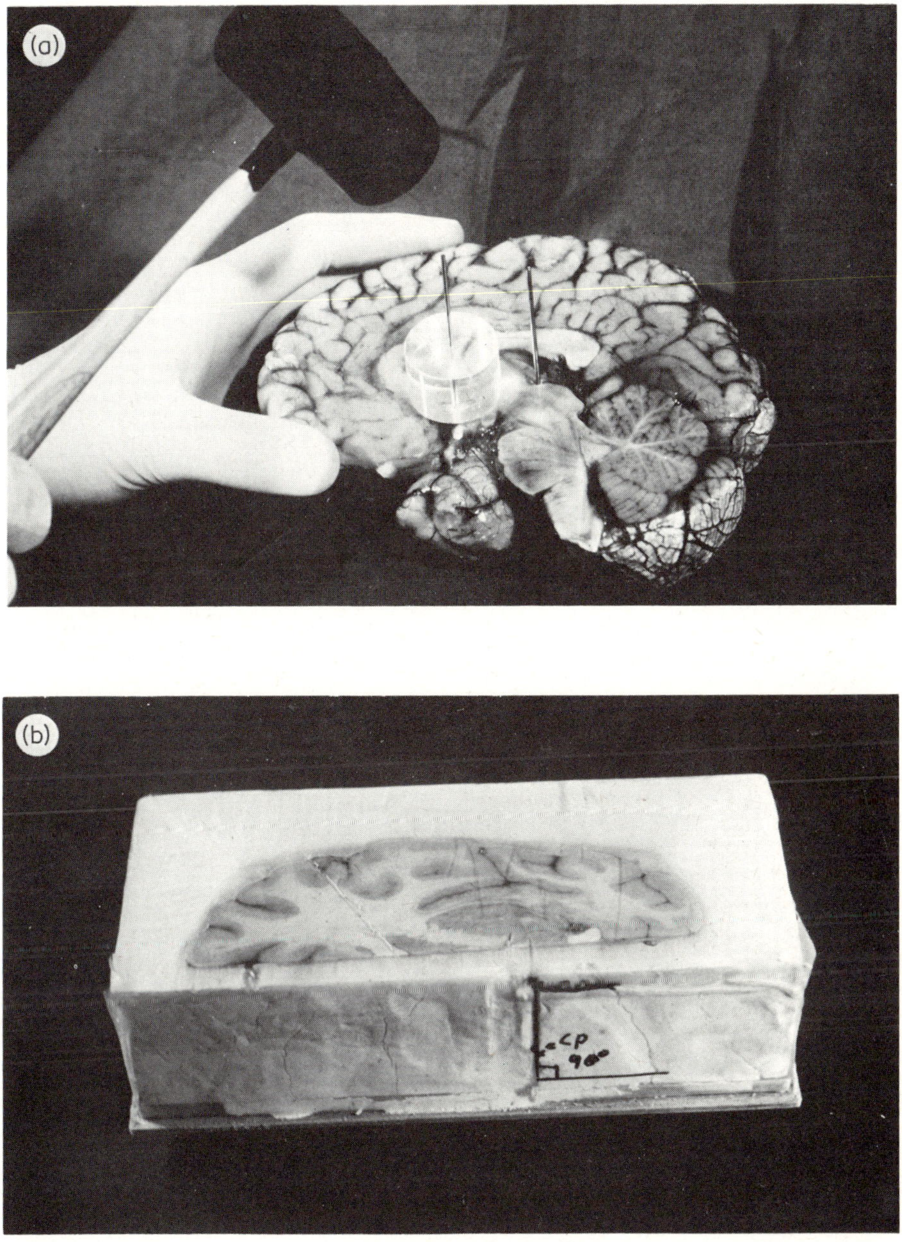

Figure 5 (a) Steel pins are introduced into the commisures perpendicular to the sagittal plane.
(b) Block including part of the hemisphere stereotaxically oriented for sectioning in the plane illustrated in Figure 9

hemisphere has earlier been worked out (Aquilonius *et al.*, 1975). After carefully dividing the brain in the midsagittal line one hemisphere is frozen with the sagittal plane facing down on a glass plate. When frozen, steel pins are introduced perpendicularly to the sagittal plane through the anterior and the posterior commisure. These steel pins are used to define a section plane perpendicular to the intercommisural and sagittal planes (Figure 5).

2.3 Sectioning

It is important to let the block equilibrate with the temperature of the cryostat (−20 °C) before sectioning begins. The specimen is cut down to the level of interest. A piece of transparent tape (3M type 800 or 810) is then fastened on

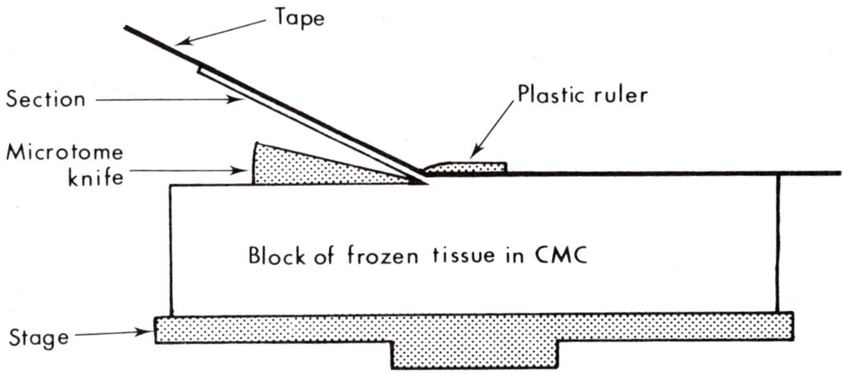

Figure 6 Schematic illustration of sectioning

to the section surface of the block by gentle rubbing. As schematically presented in Figure 6 the knife cuts beneath the tape. The transparent tape then carries the tissue section and keeps it intact (Figure 3).

Training is necessary for a good result—especially when working with large sections of a thickness of 100 μm or more. In these situations a plastic ruler is firmly pressed on the tape surface just in front of the knife.

3. PROCESSING OF SECTIONS

3.1 Collecting samples for biochemical analysis

The technique of punching out samples from cryosections has been in use for several years (Aquilonius *et al.*, 1973; Palkovits 1973; see also Chapter 4). In our laboratory the tissue section is placed on a special cryotable (LKB, Sweden) with variable surface temperature (Figure 7). For punching from nervous tissue the ideal temperature is about −10 °C. For small specimens, such as spinal cord, a stereomicroscope is necessary. Stainless steel punch-tools of different diameters have to be made from cannulas bevelled in the tips (Figure 8). Punching is performed by gentle rotation of the tool perpendicular to the surface. The tape should be left intact. When punching the tools are first cooled on dry ice. To avoid thawing and to enable transfer

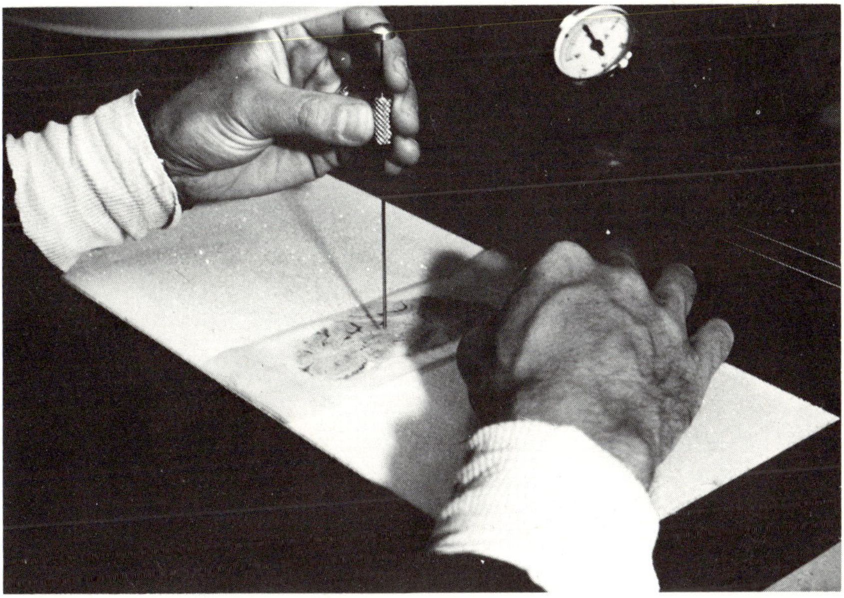

Figure 7 Punching out tissue samples on the cryotable (LKB, Sweden)

of the small tissue sample to the bottom of a micro-vial the tool is cooled again after punching and the frozen tissue is expelled by the mandril.

The punching from tape strips is a rapid method of obtaining reproducible tissue samples and biochemical data can be expressed per unit of volume (e.g. nkatal l^{-1}). The reproducibility of the punch-sampling technique was checked by weighing consecutive tissue samples on an automatic electrobalance (Cahn model 4700, USA). Using the 0.6 mm diameter tool (113 nl volume when used in 400 μm cryosections), the weights were found within narrow limits, 86.0 ± 2.5 μg (mean ± S.E., n = 12) 30 seconds after punching.

An alternative method in some situations is to cut out whole structures such as defined nuclei from the tape using a small knife. However, calculation of tissue volume is more complicated and involves an indirect step such as weighing of photos or planimetric procedures.

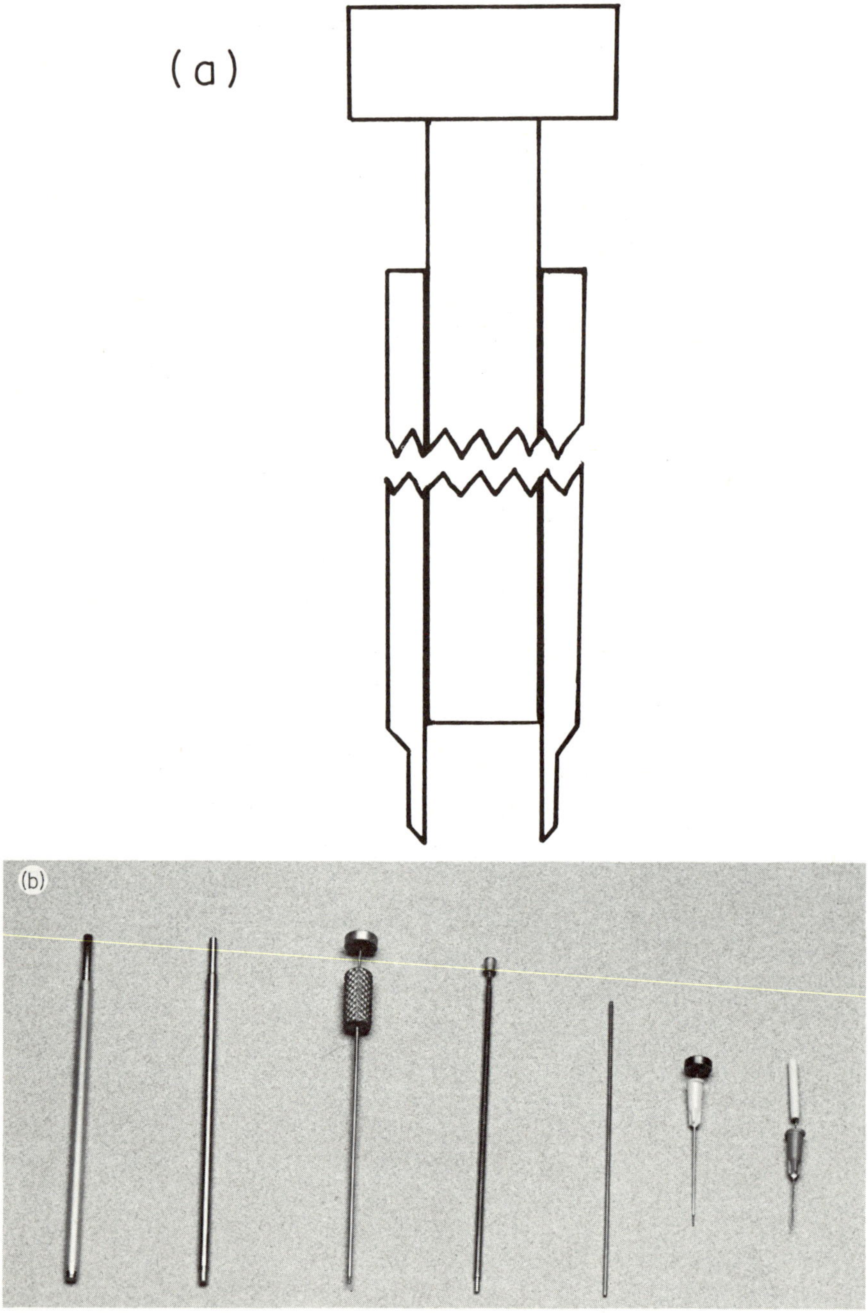

Figure 8 Schematic drawing of (a) punch tool and (b) different types of tools used in our laboratory

For mapping, larger cryosections are always photographed before and after punching while smaller sections are usually preserved fixed in 10% formaldehyde.

3.2 Notes regarding safety precautions

The handling of unfixed material from human brain and spinal cord involves a theoretical risk for contagion with infectious agents including transmissible

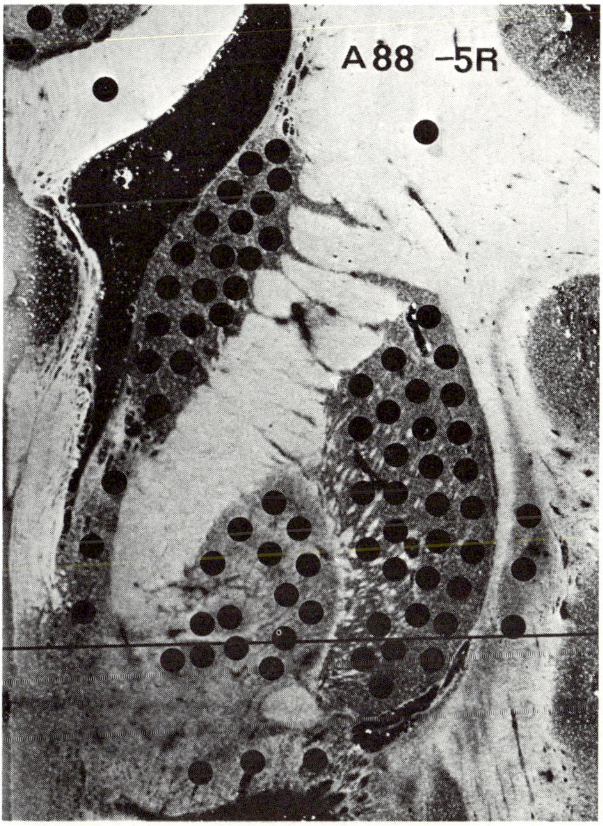

Figure 9 Distribution of tissue punches within a cryosection through extrapyramidal structures 5 mm behind the anterior commisure (Reproduced with permission from Aquilonius *et al.*, 1975)

virus dementia. Although it is difficult to establish an entirely safe protocol for cryosectioning, some strict precautions should be taken.

In our laboratory the following precautions are taken. Material from patients with suspected Creutzfeldt-Jacob disease is never used. All work is

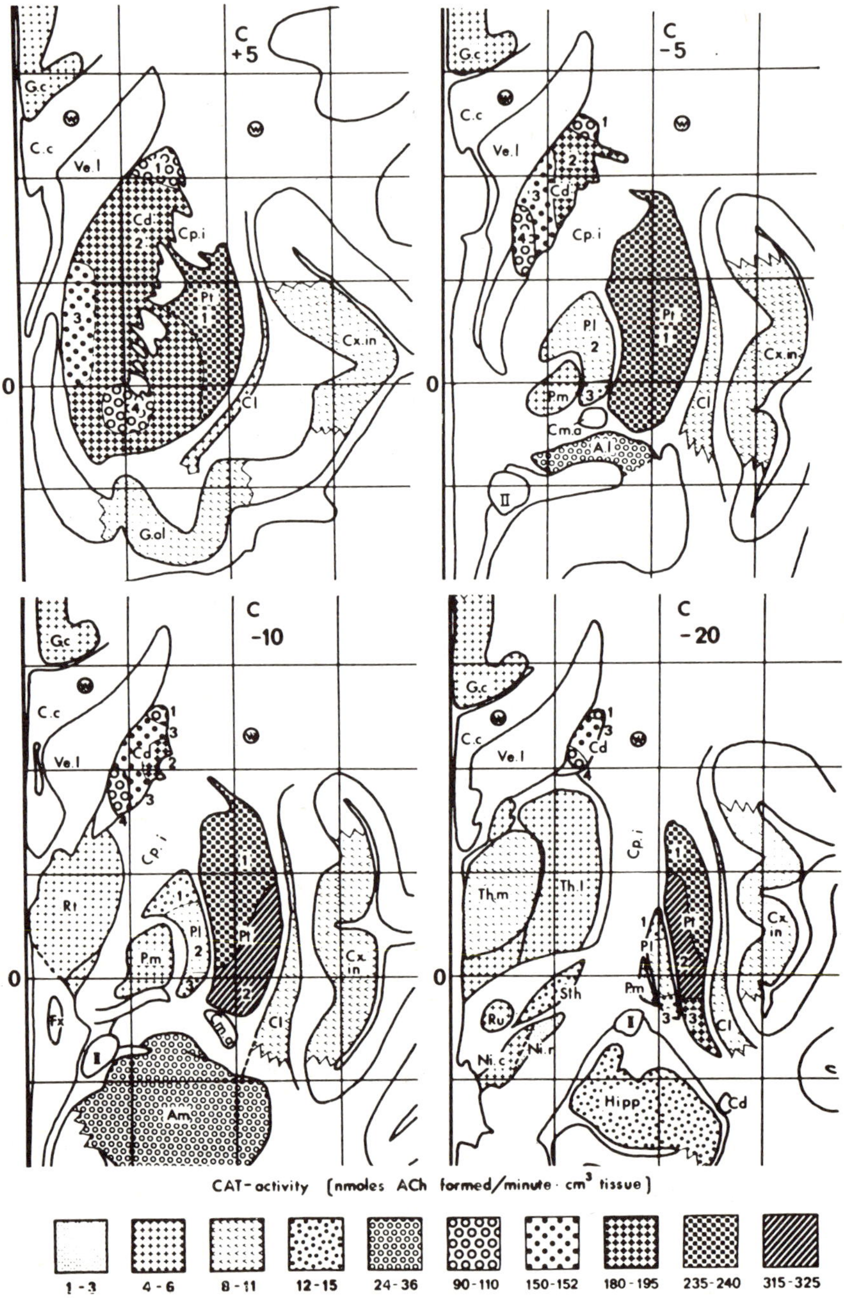

Figure 10 Choline acetyltransferase activities in different parts of the human brain (Reproduced with permission from Aquilonius *et al.*, 1975)

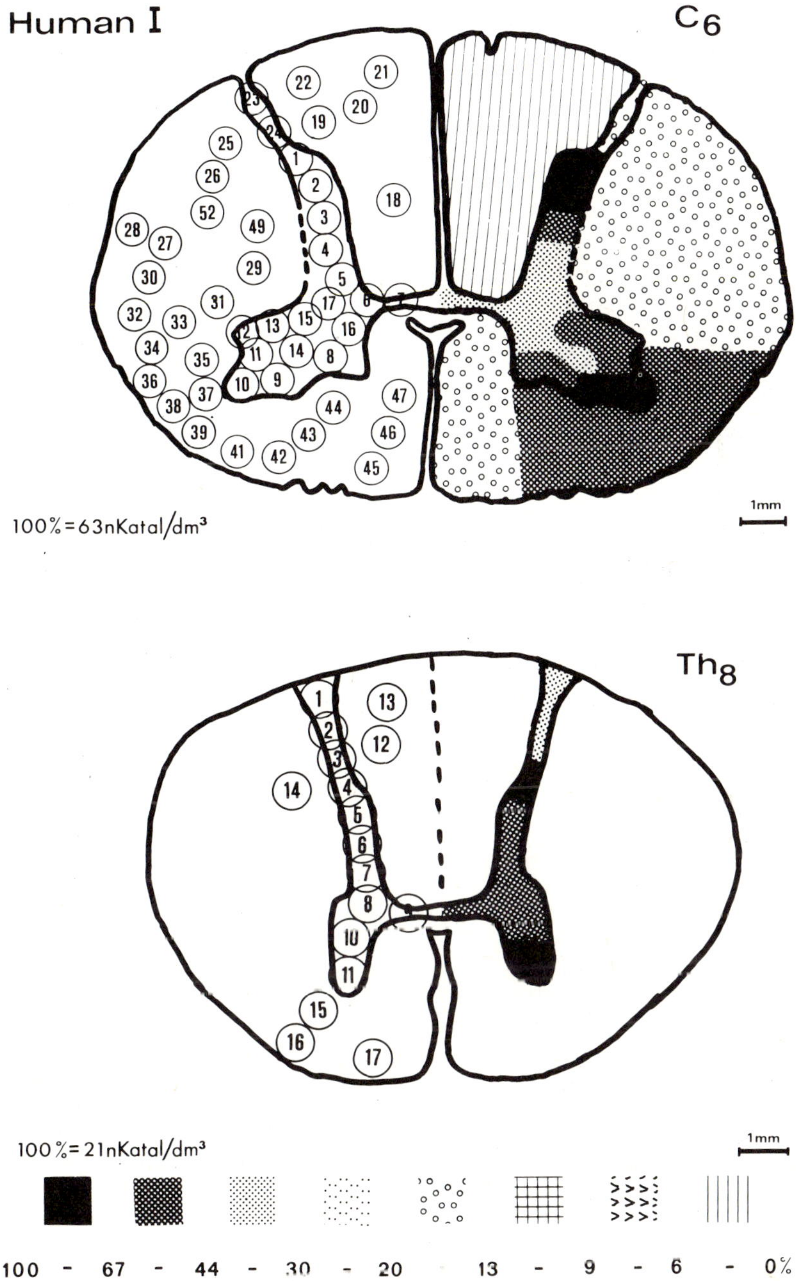

Figure 11 Choline acetyltransferase activities in the human spinal cord (Reproduced with permission from Aquilonius *et al.*, 1981)

performed within a closed laboratory. Small particles within the cryostat are removed daily by a built-in vacuum cleaner. During this procedure extra protective clothing, gloves, and a surgical mask are worn. 5% Chloramine is used for disinfection of cryotable and tools.

4. EXAMPLES OF APPLICATION

4.1 **Biochemical mapping in human brain and spinal cord**

Figure 9 demonstrates the distribution of tissue punches within a cryosection through extrapyramidal structures 5 mm behind the anterior commisure. The brain is oriented as described above with a section plane perpendicular to the intercommisural and sagittal planes.

Figure 10 illustrates one way of presenting the results (Aquilonius *et al.*, 1975). Positions with similar mean activities of the cholinergic marker choline acetyltransferase (ChAT) were grouped together and differences between neighbouring groups were tested for statistical significance with the Student's two-tailed *t*-test.

A similar example is presented in Figure 11 showing the distribution of punch-positions and mean ChAT activities within 13 cervical and four thoracic consective 400 μm sections of the human spinal cord (Aquilonius *et al.*, 1981).

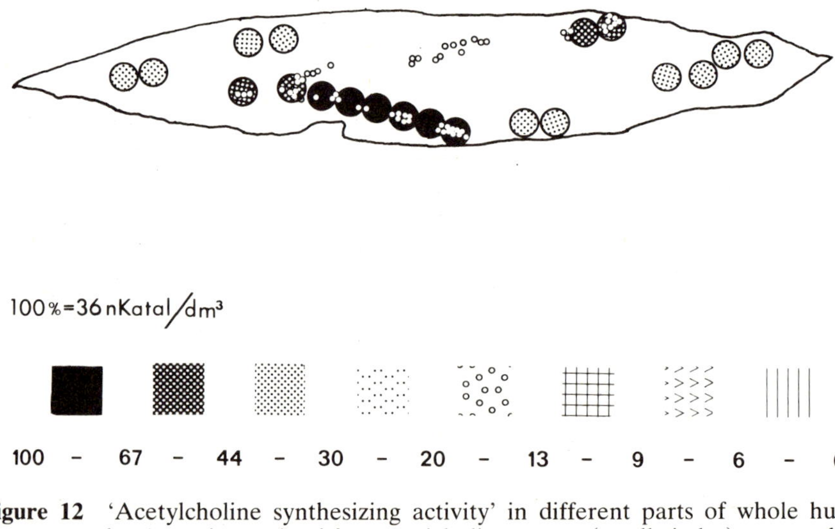

Figure 12 'Acetylcholine synthesizing activity' in different parts of whole human biceps muscle. A section stained for acetylcholinesterase (small circles) was used as a map for tissue punches (large circles) used for radioenzymatic analysis

4.2 **Regional biochemistry in whole human muscle**

In a project intended to measure ChAT-activity and nicotine-like binding sites in human muscle biopsies we found it important to get well-defined tissue samples from end-plate-rich and non-end-plate regions of human muscle. Figure 12 demonstrates the 'acetylcholine synthesizing activity' in different parts of whole human biceps muscle. Acetylcholinesterase staining in a neighbouring section was used as a guide for punching.

Large cryosections of human muscles have recently been used for histochemical and X-ray microanalytical investigations by Wroblewski *et al.*, (1980).

4.3 **Neuroanatomy**

Computed tomography has been established as a most important tool in neuroradiology. The technique is mainly restricted to near horizontal planes, a fact which creates a need for anatomical knowledge of brain sections which have not been commonly studied previously. Therefore we applied large section cryomicrotomy of whole human brain and produced a colour atlas of horizontal sections which were photographed just thawed against a black background (Aquilonius and Eckernäs 1980). After carefully removing the fresh brain from the skull, the cortex is covered with a thin plastic film and the brain is frozen within its own deep skull-cap containing about 100 ml of ethyleneglycol. In this way the frozen undeformed brain can be removed. Figure 13 demonstrates a 200 μm unstained section 70 mm below the vertex.

4.4 **Neuropathology**

Rauschning and collaborators (1980) have recently successfully combined cryomicrotomy and radiological tomography of the frozen specimen for the analysis of different pathological conditions.

Figure 14 demonstrates the possibility of comparing tomogram and cryosections of the sella turcica region (Muhr *et al.*, 1981).

A similar approach has been used in the analysis of cervical fractures with spinal cord compression (Rauschning *et al.*, 1980).

ACKNOWLEDGEMENTS

The investigations performed in the authors' laboratory were supported by the Swedish Medical Research Council (project 4373). The authors are indebted to Mrs Gun Kärrlander for secretarial help.

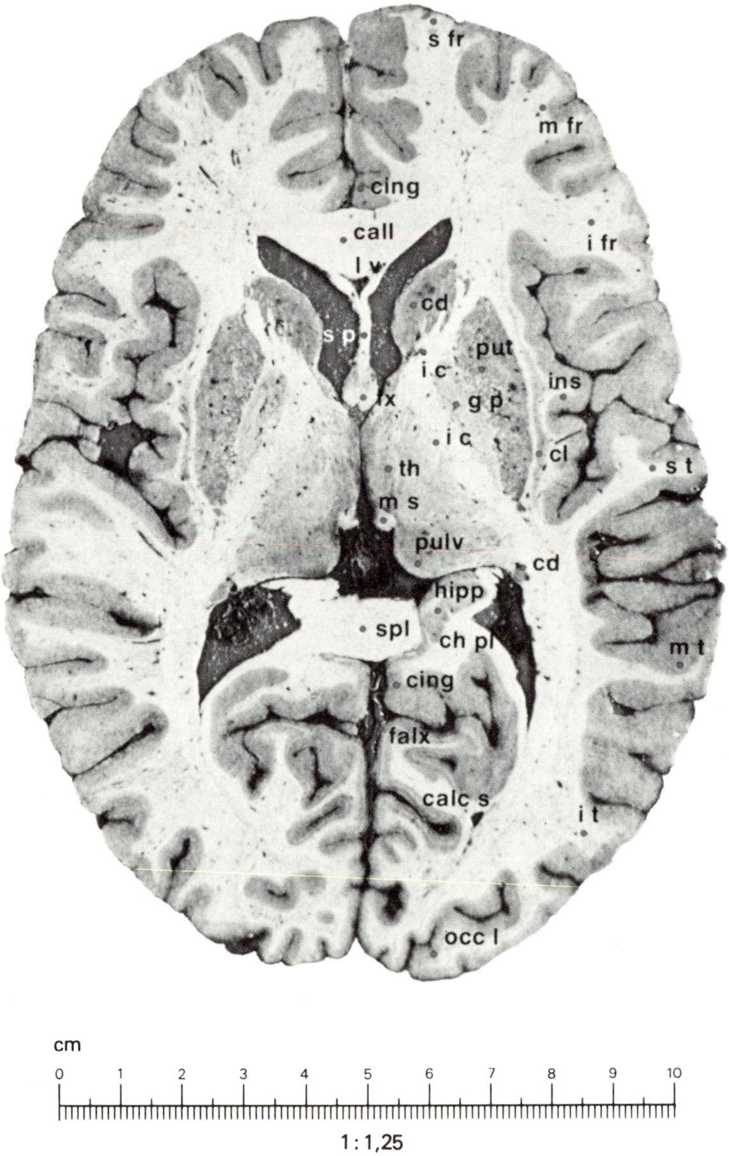
s fr
m fr
cing
call
i fr
l v
cd
s p
put
i c
ins
fx
g p
i c
cl
s t
th
m s
pulv
cd
hipp
spl
ch pl
m t
cing
falx
calc s
i t
occ l
cm
0
1
2
3
4
5
6
7
8
9
10
1 : 1,25

REFERENCES

Aquilonius, S. M., Eckernäs, S. Å., and Sundwall, A. (1973) 'Topical localization of choline acetyltransferase (ChAc) in human extrapyramidal nuclei.' *Acta. Physiol. Scand. Suppl. 396,* **Abstract 116.**

Aquilonius, S. M., Eckernäs, S. Å., and Sundwall, A. (1975) 'Regional distribution of choline acetyltransferase in the human brain: Changes in Huntington's chorea'. *J. Neurol. Neurosurg. Psychiat.*, **38,** 669–677.

Aquilonius, S. M., and Eckernäs, S. Å. (1980) *A colour atlas of the human brain adapted to computed tomography.* Esselte Studium, Stockholm.

Aquilonius, S. M., Eckernäs, S. Å., and Gillberg, P. G. (1981) 'Topographical localization of choline acetyltransferase within the human spinal cord and a comparison with some other species.' *Brain Res.* **211,** 329–340.

Kopp, N., Denoroy, L., Renaud, B., Pujol, J. F., Tabib, A., and Tommasi, M. (1979) Distribution of adrenaline-synthesizing enzyme activity in the human brain.' *J. Neurol. Sci,* **41,** 397–409.

Muhr, C., Bergström, K., Larsson, S.-G., and Rauschning, W. (1982) 'The sella turcica and the pituitary gland: a correlative anatomic study with cryomicrotomy and skeletal tomography in 40 autopsy specimens.' *Acta Radiol.*, (in press).

Palkovits, M. (1973) 'Isolated removal of hypothalamic or other brain nuclei of the rat'. *Brain Res,* **59,** 449–450.

Rauschning, W. (1980) 'Anatomy and function of the communication between knee joint and popliteal bursae.' *Ann. Rheumatic Dis.*, **39,** 354–358.

Rauschning, W., Sahlstedt, B., and Wigren, A (1980) 'Irreponible Luxationsfraktur der unteren Halswirbelsäule. Eine pathologisch-anatomische studie mit der Gefrierschneidemethode.' *Chirurg.*, **51,** 529–533.

Ullberg, S. (1978) 'The technique of whole body autoradiography: Cryosectioning of large specimens.' *Science Tools,* **1977**.

Wroblewski, R., Edström, L., and Rauschning, W. (1980) 'Histochemical and X-ray microanalytical investigations of defined organs and cells in their true position within whole-body cryosections.' *Histochem.*, **67,** 43–52.

Figure 13 Cryosection in the canto-meatal plane through a whole human brain 70 mm below the vertex. The photograph of the just tawed 200 μm section was originally taken in colour against a black background (Reproduced with permission from Aquilonius and Eckernäs, 1980). Abbreviations: calcarine sulcus, calc s; caudate nucleus, cd; choroid plexus, chpl; cingulate gyrus, cing; claustrum, cl; corpus callosum, call; falx cerebri, falx; fornix, fx; globus pallidus, gp; hippocampus, hipp; inferior frontal gyrus, ifr; inferior temporal gyrus, it; internal capsule, ic; insula cortex, ins; lateral ventricle, lv, medullary stria, ms; middle frontal gyrus, m fr; middle temporal gyrus, mt; occipital lobe, occl; pulvinar, pulv; putamen, put; septum pellucidum, sp; splenium, spl; superior frontal gyrus, sfr; superior temporal gyrus, st; thalamus, th

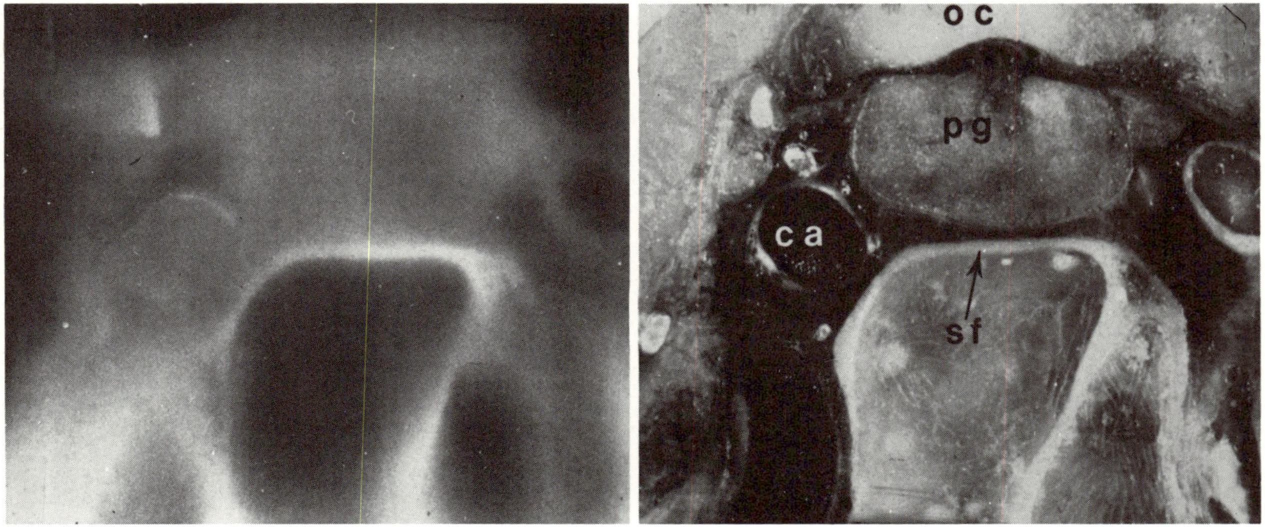

Figure 14 X-ray tomogram (left) and cryosection (right) in the same frontal plane through the sella turcica region. Abbreviations; carotid artery, ca; optic chiasm, oc; pituitary gland, pg; sellar floor, sf. (The figure was kindly provided by Dr Carin Muhr)

Brain Microdissection Techniques
Edited by A. C. Cuello

CHAPTER 5

Microdissections of Nuclear Areas from Human Brain Slices

THOMAS J. STEVENS and EDWARD D. BIRD

Brain Tissue Resource Center, Ralph Lowell Laboratories, McLean Hospital, Harvard Medical School Belmont, Massachusetts 02178 USA

CONTENTS

1. General comments on the sampling of anatomically defined areas from the human brain 171
2. Handling of the brain 172
3. Preparation 173
4. Cutting 173
5. Dissection 173
6. Criteria used for identification of areas 173
   6.1 Area 10 177
   6.2 Areas 8 and 9 177
   6.3 Area 11 178
   6.4 Area 12 178
   6.5 Area 32 179

1. GENERAL COMMENTS ON THE SAMPLING OF ANATOMICALLY DEFINED AREAS FROM THE HUMAN BRAIN

Neurochemical studies on post-mortem human brain tissues have demonstrated chemical stability for a number of neurotransmitters (Spokes, 1980), their biosynthetic enzymes (Bird and Iversen, 1974), and receptor binding for labelled ligands (Enna *et al.*, 1976) provided the corpse has been cooled promptly after death and the brain frozen upon removal from the skull. Significant neurochemical abnormalities have been found in post-mortem brain of Parkinson's disease (Agid and Agid, 1980; Eringer and Hornykiewicz, 1960), Huntington's chorea (Bird, 1980) and Alzhiemer's disease (Bowen *et al.*, 1976; Davies and Maloney, 1976; Perry *et al.*, 1977), in specific areas of the human brain and, therefore, an accurate dissection of the brain is required to distinguish these differences.

Certain neurological diseases are relatively rare, and in order to collect enough cases to make a statistical comparison with normal brain, a common resource for the collection of such brains is usually necessary. Since post-mortem specimens are often obtained in locations far removed from the neurochemical laboratory, freezing the tissues ensures biochemical stability and consistency between specimens with regard to post-mortem variables. One important variable could be the dissection of both the abnormal and normal brain, and this Chapter will deal with a few important features of handling and dissecting human brain.

The technique of dissecting human brain presented here varies markedly in both method and magnitude from other techniques presented in this series. The methods utilized by the authors are perhaps properly classified as 'macrodissection' in contrast to the relative 'microdissection' so often described for small animals, using the punch, microtome, and microscopic dissection techniques (see Chapters 1, 2, and 3). Each method presents its own advantages, as well as limitations.

Each human brain is unique and less standardized than those of small in-bred animals. Atlases may assist in identifying the anatomical relationship between human nuclei, but there is no convenient uniformity as shown in animals such as the rat, where a standard atlas and its stereotaxic co-ordinates may be used to define a nucleus.

Perfusion, control of diet or medications, decapitation, and other standard laboratory procedures often used in animals are not possible in human subjects. Such variables as the time lapse between death and the time of specimen removal, medications, and diet can be documented, but rarely controlled prior to the subject's death.

Due to the less uniform nature of human anatomy and the agonal effects at death, it is important to standardize the post-mortem conditions for the collection of brain tissues to allow comparison between individual cases. Dissection techniques also need be standardized. Researchers must be able to compare data gathered in analogous areas in a series of cases against data generated by other investigators. The need for uniform handling and dissection techniques is the basis for having established a brain tissue resource centre.

The brain tissue bank that is currently operating at Harvard Medical School's McLean Hospital in Belmont, Massachusetts provides a large number of uniformly handled and dissected samples from each brain to many independent researchers, promoting more research and less waste from a given specimen than would otherwise occur.

2. HANDLING OF THE BRAIN

When the brain of a deceased subject is to be donated, the pathologist should carry out the autopsy promptly and immediately remove the brain from the skull, cut it in the midline sagitally and place one hemisphere in a fixative (usually 10% buffered formalin) for histopathological examination, and the other hemisphere is frozen at −20 °C. The frozen specimen is shipped in an insulated container filled with dry-ice to the resource centre, where it is then stored at −70 °C. Biochemical stability for a number of neurochemicals (E. D. Bird, unpublished results) has been shown in tissue stored under these conditions for a number of years.

3. PREPARATION

Prior to dissection of a frozen hemisphere, a neuropathology report should be completed and the diagnosis established. The specimen may then be removed from the −70 °C freezer to a −20 °C unit 12–24 hours prior to cutting so as to allow sections to be easily and uniformly cut. Tubes are then labelled with the case number and coded for the various areas which will be dissected. A typical sample vial has a screw top with a silicone gasket to prevent dehydration or accidental loss of tissue.

4. CUTTING

A U.S.–Berkel commercial meat slicer with a custom serrated blade is used to cut 4 mm thick coronal sections from the frozen brain. Each slice is wiped clean of frothy residue created by both the heat of friction from the blade and its kerf. The slice is then placed on a −10 °C mechanically cooled cold plate. The tissue, in this way, is never in danger of thawing.

5. DISSECTION

All areas dissected are identified visually according to stereotaxic atlases (Olszewski and Baxter, 1954; Riley, 1969; Schaltenbrand and Wahren, 1977) and Brodmann areas (Brodmann, 1909; von Bonin, 1960), for cerebral cortex, (Figure 1). Pre-cooled scapels (no. 10 and no. 11 blades) are used with pre-cooled forceps to remove nuclei. When the dissector is satisfied that all tissue of a given area has been removed, the accumulation of tissue is finely chopped and mixed manually to create a homogeneous frozen mass. The areas dissected in this fashion are listed in Table 1, and tissue from such areas consists of grey matter only, reasonably free from any blood vessels, dura mater, or white matter. The latter is collected and stored in large scintillation vials. An aliquot from each area is placed into pre-cooled, labelled vials, which are placed on dry-ice pending storage in a −70 °C freezer.

6. CRITERIA USED FOR IDENTIFICATION OF AREAS

Identification of cortical areas and many nuclei is accomplished visually with reference to human atlases as well as reliance upon experience. Some areas,

Table 1 Areas routinely dissected

- Basal ganglia
  - I. Caudate
  - II. Lenticular nuclei
    - A. Putamen
    - B. Globus pallidus
      - 1. Inner (medial)
      - 2. Outer (lateral)
- Basal pons
- Brain stem nuclei
  - I. Central grey
  - II. Dorsal tegmental area
  - III. Locus coeruleus
  - IV. Olive
  - V. Red nucleus
  - VI. Subthalamic nucleus
  - VII. Substantia nigra
    - A. Zona compacta
    - B. Zona reticulata
  - VIII. Ventral tegmental area
- Bed Nucleus of stria terminalis
- Broca's Area (see cerebral cortex–Brodmann Area 44)
- Cerebellum
  - I. Cerebellar cortex
  - II. Dentate nucleus
- Cerebral Cortex
  - Brodmann Areas 4,6,8,9,10,11,12,17,22,24,25,28,32,38, and 44
- Hypothalamus
- Thalamus
  - I. Anterolateral nucleus
  - II. Anteromedial nucleus
  - III. Dorsalmedial nucleus
  - IV. Ventrallateral nucleus
- Pineal
- Limbic nuclei
  - I. Amygdala
  - II. Hippocampus
  - III. Nucleus accumbens
  - IV. Septal nuclei

however, are more ambiguous and require adherence to prescribed methods,

The nucleus accumbens, for example, has been shown by our preliminary, unpublished data to consist of biochemically heterogeneous tissue. The anatomical limits of this area have long been disputed, but in September, 1980, a group of neuroscientists who dissect human brain gathered in Cambridge, U.K. to discuss standardization of dissection techniques for the human brain. A sub-committee was established to discuss the nucleus accumbens and an agreement was reached to define and dissect the nucleus according to the criteria established by von Brockhaus (1942).

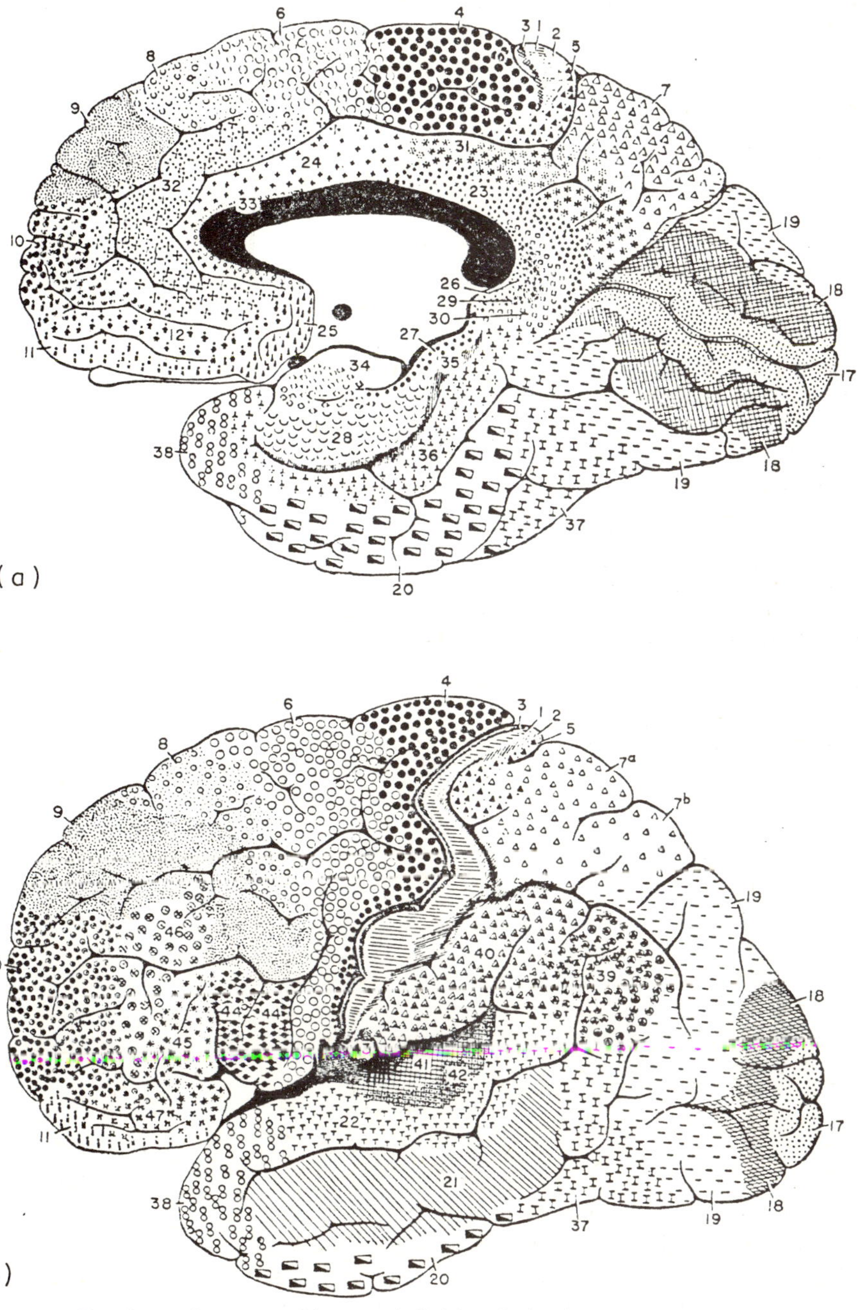

Figure 1 Brodmann's cytoarchitectural fields of the human cerebral cortex. (a) Medial view. (b) Lateral view. (Reproduced with permission from Crosby *et al.*, 1962. *Correlative Anatomy of the Nervous System*. Copyright © 1962 by Macmillan Publishing Co., Inc)

Under the 'new' definition by the group of neuroscientists, the accumbens consists of the caudate/putamen formation that lies anterior of the anterior commisure, posterior to the point at which the internal capsule reaches its most ventromedial aspect, and ventral to a diagonal cut across the caudate, internal capsule, and putamen, starting at the level on the medial aspect at which the lateral ventricle reaches its most ventral point, and proceeding at an angle of approximately 45 ° ventrolaterally (see Figure 5).

It is beyond the scope of this Chapter to describe the anatomical limits of each area dissected. What is important to remember is that our function as a brain tissue resource centre is to provide as many individual tissue samples as possible from a given brain specimen with homogeneity between samples of any one given area of a single particular specimen. Our methods, which are gross in comparison to the various, finer microdissection methods reported in this handbook, rely on visually identifying and removing areas of interest, pooling and mixing finely all similar tissue to create a uniform, homogeneous, frozen mass from which individual, representative samples are aliquoted.

The maintenance of the tissue in its frozen state has been described and justified earlier in this Chapter. Although many areas are heterogenous, comparison of data and replication of results would not be possible if there were not homogeneity between samples identified under a single classification. The authors have begun the study of several areas, dividing each into progressively smaller units so as to obtain a more specific and less regional measurement of biochemical activity, but, at the same time, we continue to dissect areas as previously defined for our tissue bank as such definitions are now well-established and represent the basis of many assays done by various researchers who utilized tissue dissected by such definitions.

The limits of areas of interest are generally well-defined in coronal sections by various authors (Olszewski and Baxter, 1954; Riley, 1960; Schaltenbrand and Wahren, 1977) with the exception of the cortical areas, thereby warranting their description in this Chapter.

Brodmann (1909) described 47 cortical fields of the human brain based upon the differences in the arrangement of cell types and the pattern of myelinated fibres revealed through his histological survey of homotypical cortex, i.e., that in which the six layers are easily distinguished. The system employed by Brodmann to number each area does not imply any significance or relationship of one area to another, and is generally based upon the order in which each area was studied. Subsequent investigation has revealed that areas previously assumed to be homogeneous were actually heterotypical, von Economo (1929). Because of these findings and the methods originally used, Brodmann's descriptions have been subject to criticism. The Cambridge meeting (A. Pope, unpublished) discussed dissection of cerebral cortex, and concluded that for the time being Brodmann's nomenclature to identify areas would be employed.

The authors use Brodmann's descriptions to identify areas of cortex visually on coronal sections of brain and are marked on the photographs shown in Figures 2–7. The marked areas represent cortical tissue that is well-removed from the boundaries of adjacent cortical areas.

6.1 **Area 10**

When slicing coronally from anterior to posterior, the first Brodmann area encountered is area 10, at the tip of the frontal lobe. As Figures 1 and 2

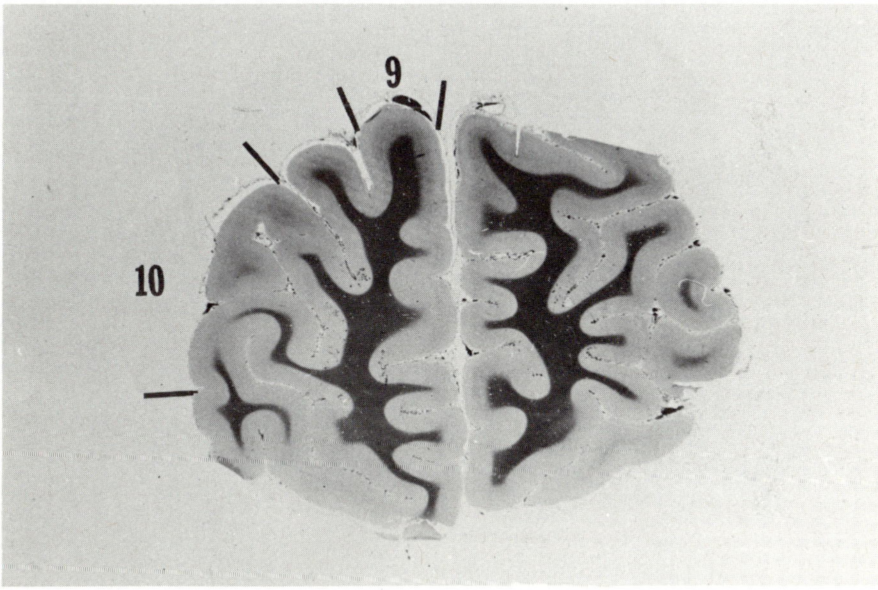

Figure 2 A coronal section approximately 8.0 mm from the tip of the frontal pole. Numbers refer to corresponding Brodmann areas

illustrate, area 10 is present on both the medial and lateral surfaces of the hemisphere. It is important to be aware that frequently the extent of a given area is not symmetric with respect to the lateral versus medial aspects.

6.2 **Areas 8 and 9**

Area 9 is seen on both the lateral and medial surface of the second slice (approximately 8 mm deep) at its most dorsal aspect, but in succeeding slices area 9 is found more ventrally on these surfaces, since area 8 first appears at the dorsal aspect on these slices. Similarly, proceeding posteriorly area 6 next appears at the most dorsal position and subsequently area 4.

6.3 **Area 11**

Within a few millimetres of the anterior tip of the frontal lobe, area 11, the olfactory cortex, begins ventrally and soon assumes its characteristic shape,

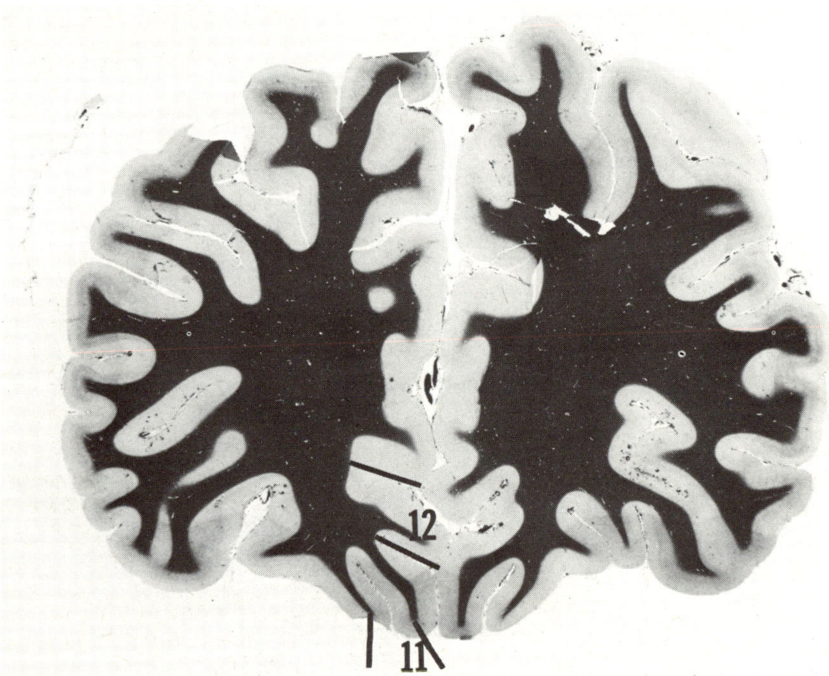

Figure 3 A coronal section approximately 25.1 mm from the tip of the frontal pole. Numbers refer to corresponding Brodmann areas

appearing enclosed in an inverted horseshoe of white matter on the orbital surface of the hemispheres, cradling the olfactory tract.

6.4 **Area 12**

Immediately dorsal to area 11, on the medial surface, Brodmann area 12 is found (Figure 3).

6.5 **Area 32**

Area 32 appears on the medial surface of the hemisphere as a midparietal gyrus between area 12 ventrally and area 9 superiorly which extends in more posterior slices to be adjacent to the cingulate gyrus.

As neurochemical interests develop in other cortical areas, these dissection techniques can be applied. In the future, when more neuronal pathways are discovered, what has been described above as macrodissection will evolve into microdissection involving thinner slices with finer dissection criteria.

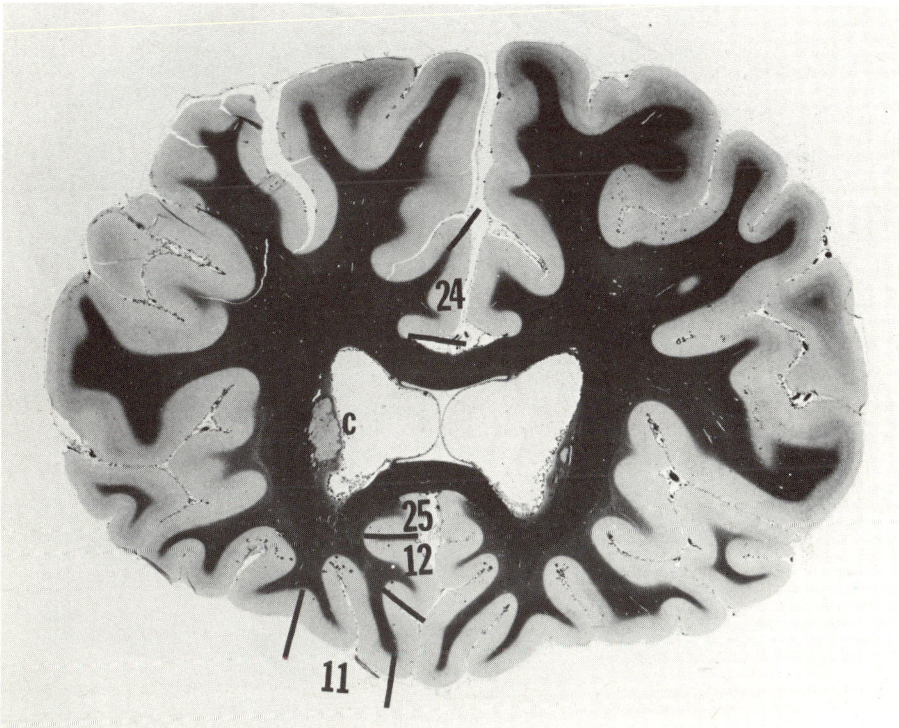

Figure 4 A coronal section approximately 31.4 mm from the tip of the frontal pole. Numbers refer to corresponding Brodmann areas caudate, c

Figure 5 A coronal section approximately 34.8 mm from the tip of the frontal pole. The black (myelinated) tissue separating the caudate, c, and putamen, p, is the internal capsule; nucleus accumbens, acc

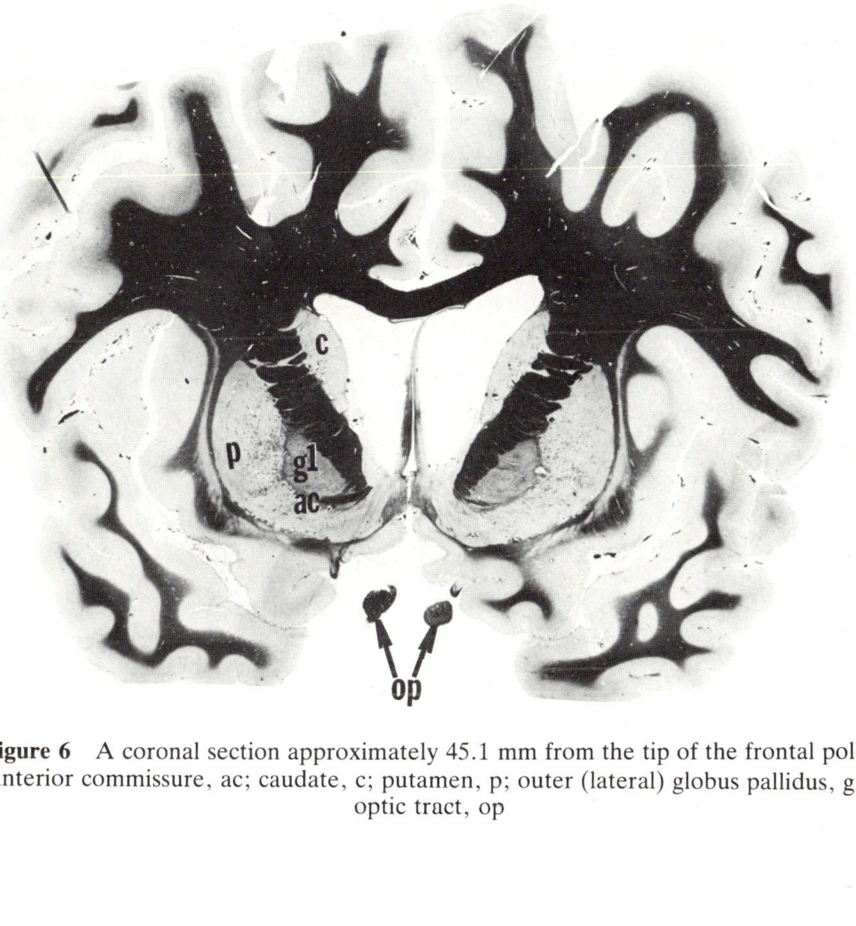

Figure 6 A coronal section approximately 45.1 mm from the tip of the frontal pole. Anterior commissure, ac; caudate, c; putamen, p; outer (lateral) globus pallidus, g1; optic tract, op

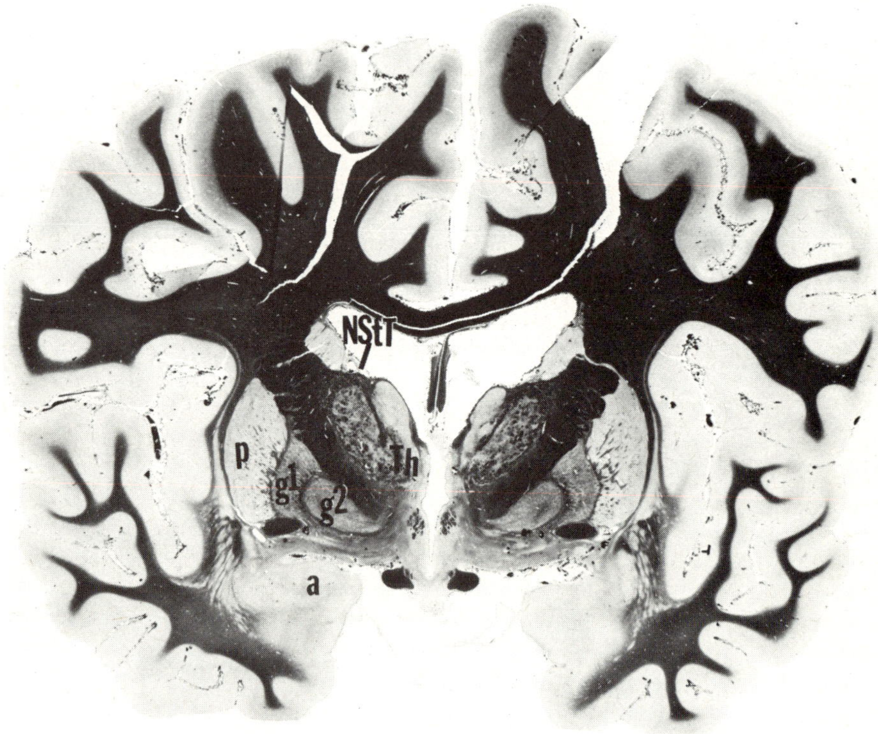

Figure 7 A coronal section approximately 49.9 mm from the tip of the frontal lobe. Amygdala, a; outer (lateral) globus pallidus, g1; inner (medial) globus pallidus, g2; bed nucleus of the stria terminapis, N st T

ACKNOWLEDGEMENT

The authors wish to acknowledge the generous assistance of Mr Peter Paskevich and Dr Louis Stinus with respect to the technical photography.

REFERENCES

Agid, F. J., and Agid, Y. (1980) 'Is the mesocortical dopaminergic system involved in Parkinson's disease?' *Neurology*, **30,** 1326–1330.

Bird, E. D. (1980) 'Chemical pathology of Huntington's disease.' *Ann. Rev. Pharmacol. Toxicol.*, **20,** 533–551.

Bird, E. D., and Iversen, L. L. (1974) 'Huntington's chorea: post-mortem measurement of glutamic acid decarboxylase, choline acetyltransferase and dopamine in basal ganglia.' *Brain*, **97,** 457–472.

Bowen, D. M., Smith, C. B., White, P., and Davison, A. N. (1976) 'Neurotransmitter-related enzymes and indices of hypoxia in senile dementia.' *Brain*, **99,** 459–496.

Brodmann, K. (1909) *Vergleichende Lokalisationlehre der Grosshirnrinde in ihren Prinzipien dargestellt auf Grund des Zellenbaues*, Barth, Leipzig.

Crosby, E. C., Humphrey, T., and Laver, E. (1962) *Correlative anatomy of the nervous system*, Macmillan, New York.

Davies, P., and Maloney, A. J. R. (1976) 'Selective loss of central cholinergic neurons in Alzheimer's disease.' *Lancet*, **ii,** 1403.

Enna, S. J., Bird, E. D., Bennett, J. P., Bylund, D. B., Yamamura, H. I., Iversen, L. L., and Snyder, S. H. (1976) 'Huntington's chorea: changes in neurotransmitter receptors in the brain.' *New Engl. J. Med.*, **294,** 1305–1309.

Ehringer, H., and Hornykiewicz, O. (1960) 'Verteilung von Noradrenalin und Dopamin (3-hydroxytryomin) im gehim des Meschen und ihr verhalten bei herkrankundgen des Extrapyramidalen Systems.' *Klin. Wocherschr*, **38,** 1236–1239.

Olszewski, J., and Baxter, D. (1954) *Cytoarchitecture of the Human Brain Stem*, Lippincott, Philadelphia.

Perry, E. K., Gibson, P. N., Blessed, G., Perry, R. H., and Tomlinson, B. E. (1977) 'Neurotransmitter enzyme abnormalities in senile dementia.' *J. Neurol. Sci.*, **34,** 247–265.

Riley, H. A. (1960) *An atlas of the basal ganglia, brain stem and spinal cord*, Hafner, New York.

Schaltenbrand, G., and Wahren, W. (1977) *Atlas for stereotaxy of the human brain*, Georg Thieme, Stuttgart.

Spokes, E. G. S. (1980) 'Neurochemical alterations in Huntington's disease.' *Brain*, **103,** 178–210.

von Bonin, G. (1960) *Some papers on the cerebral cortex*, Thomas, Springfield, Illinois.

von Brockhaus, H. (1942) 'Zur feineren Anatomie des Septum und des Striatum.' *Journal Fur Psychologie Und Neurologie*, **51,** 1–55.

von Economo (1929) 'The cytoarchitectonics of the human cerebral cortex.' Oxford University Press, London.

Subject Index

accumbens, 112, 123, 174, 176
acetylcholine, 3, 127, 166
acetylcholinesterase, 128, 167
aliquots (aliquoted), 28, 33, 150, 173, 176
amino acids, 124, 128, 129, 139, 150
amygdala (amygdaloid, amygdaloideus), 20, 120, 123
ante-mortem, 129
anterior commissure, 21, 30, 112, 166, 176
arcuate nucleus, 12

basal ganglia, 128
biochemical, 2, 13, 27, 29, 32, 33, 49, 122, 123, 128, 148, 150, 160, 166, 172, 173, 174, 176
blood vessels, 173
brain nuclei, 7, 9
brain stem, 5, 39, 41, 112, 124
bregma suture, 4

calvarium, 4
caudate nucleus, 9, 13, 21, 22, 30, 45, 112, 134, 176
cell bodies, 50, 123
cerebellar peduncles, 113, 133
cerebellum, 13, 16, 39, 41, 127, 129, 130, 132, 133
cerebral cortex, 16, 34, 128, 173, 175, 176
choline acetyltransferase, 50, 127, 164, 165, 166, 167
cold box, 139, 140, 141, 144, 146, 147, 148, 149
cold room, 38, 122, 129, 134
coronal plane, 5, 23
corpus callosum, 21, 22, 30, 31, 112
cortex, 13, 50, 167
cranial vault, 4, 5, 6
crus cerebri, 112, 118
cryostat, 6, 9, 10, 11, 13, 34, 117, 118, 122, 124, 129, 134, 135, 136, 137, 139, 151, 156, 157, 160, 166
cryotable, 161, 166
cuneatus, 112

dentate gyrus, 16
diencephalic, 112
dopamine, 45, 123, 124, 128, 129
dura mater, 173

electron microscopy, 2, 12, 16, 32, 122, 127, 128
enzyme(s), 2, 37, 122, 127, 128, 129, 134, 150, 171
epi-illumination, 19

fasciculus gracilis, 112
fibres, 29, 112, 176
foramen magnum, 4
formalin, 21, 30
fresh (tissue slices, brain slices), 2, 5, 6, 7, 20, 29, 37, 39, 40, 45, 46, 49, 50, 51, 113, 117, 118, 121, 122, 123, 124, 133
frozen sections, 5, 6, 7, 9, 13, 20, 23, 27, 29, 30, 31, 33, 34

GABA (γ-aminobutyric acid), 124, 127, 128, 129, 139
GAD (glutamic acid decarboxylase), 50, 127, 129
globus pallidus, 24, 112
gracilis, 113
grey matter, 9, 30, 123, 173
grid microdissection, 137, 139, 150

habenula, 112
hippocampus, 16, 26
histochemical, 128, 139
homogenization (homogenizers), 22, 25, 27, 29, 32, 33, 148, 150
hypoglossal, 113, 120
hypothalamus, 4, 12, 34, 112, 113, 123, 130

immunohistochemistry, 127, 128, 139
interstitial nucleus, 21, 22, 30, 112
irradiated brains (microwave), 2, 6, 11, 12

knife, 7, 10, 11, 12, 13, 16, 21, 25, 31, 49, 51, 129, 130, 146, 156, 160, 161
knife holder, 156

landmarks, 11, 19, 22, 23, 27, 34, 49, 50, 112, 120, 122, 123, 129, 150
layers (*or* layered), 16, 50, 120, 121
limbic system, 34
loupe, 8, 13, 14

mamillary body, 7, 129
median eminence, 12, 112, 113
medulla oblongata, 35, 112, 113, 120
meninges, 6, 39
mesencephalon, 35, 50, 112, 117, 118
microphotography, 46, 49, 51
microscale, 144, 145, 146, 147, 151
microtome, 5, 9, 10, 21, 31, 134, 137, 154, 157, 172
microwave (irradiated brains), 2, 6, 11, 12

needles, 13, 15, 16, 17, 18, 19, 20, 21, 23, 24, 25, 26, 27, 28, 29, 38, 41, 47, 48, 49, 51, 150
nerve terminals, 123
neuropeptides, 128
neurotransmitter, 3, 37, 45, 49, 124, 171
nigro-striatal, 128
nucleus solitarius, 112
nucleus spinalis, 113

occipital muscles, 4
olfactory bulb, 3, 16, 39, 46, 120, 121, 124
olfactory tubercle, 7, 16
optic chiasm, 22
optic nerves, 4, 5, 39

pars compacta, 117, 118, 123
pars lateralis, 117, 118
pars reticulata, 112, 117, 118
periventricular nuclei, 25, 27
pons, 8, 35, 113, 124
post-mortem, 2, 3, 23, 28, 32, 38, 129, 171, 172
punch (or punching), 2, 3, 6, 7, 13, 15, 16, 17, 18, 19, 20, 21, 22, 23, 24, 25, 27, 28, 29, 30, 31, 32, 33, 34, 35, 37, 156, 160, 161, 162, 163, 166, 167, 172

release, 45, 49, 122, 124
reticular, 112
reticulata, 123

scalpels, 173
sella turcica, 167
serotonin (5HT), 124
stain (*or* stained sections), 29, 30, 31, 37, 50, 122, 123, 137, 166
stereomicroscope, 8, 13, 14, 15, 29, 47, 48, 49, 161
stereotaxic, 50, 51, 122, 159, 172, 173
stria terminalis, 21, 22, 30, 112
substantia gelatinosa, 113
substantia nigra, 9, 45, 112, 113, 117, 118, 123, 124, 127, 128, 129, 130
superior colliculus, 16
suprachiasmatic, 112

telencephalic, 112
thalamus, 34
tissue bank, 176
tissue chopper, 8, 9, 39, 41, 45, 46, 50, 112, 122, 124
tissue culture, 2, 29, 32
tractus solitarius, 113, 120
trans-illumination, 14, 38, 46, 47, 48, 49, 50, 112, 122, 123, 124
trigeminal, 5, 39, 112, 113
trimming, 134
tyrosine hydroxylase, 50, 127, 129

uptake, 45, 49, 122

vagus, 113, 120
ventricles, 20, 21, 22, 23, 30, 38, 120, 123, 176
ventromedialis, 123
Vibratome, 39, 44, 45, 46, 50, 122, 124

weighing, 150, 161
white matter, 22, 23, 30, 173, 178